SxI - Springer for Innovation / SxI - Springer per l'Innovazione

Volume 18

SxI – Springer per l'Innovazione: L'Innovazione riveste nella società attuale un ruolo di fondamentale importanza, sia a livello di ricerca sia di implementazione in ambito scientifico, tecnologico, organizzativo, finanziario, legale, economico e commerciale. Springer, con il lancio di questa serie, intende proporre una serie di contributi, italiani e stranieri, nati da realtà che sviluppano, in vari settori e da diversi punti di vista, aspetti ed esperienze innovative. I volumi risultano essere uno strumento di approccio e approfondimento per tutti coloro che lavorano per la ricerca e l'innovazione, siano essi in ambito accademico, istituzionale o industriale.

SxI - Springer for Innovation: Innovation plays a fundamental role in our Society, at a research and also at an implementation level in the scientific, technological, financial, legal, economical and commercial areas. Springer, launching this series, aims at offering a wide range of contributions born in many – Italian and International - fields which develop, from different points of view, innovative aspects and experiences. The works included in this series are a useful tool meant to approach and deepen these aspects for all those who are involved in research and innovation, at an academic, institutional and industrial level.

Matteo Mazziotta · Adriano Pareto

Statistics for Composite Indicators

Springer

Matteo Mazziotta
Italian National Institute of Statistics
Rome, Italy

Adriano Pareto
Italian National Institute of Statistics
Rome, Italy

ISSN 2239-2688 ISSN 2239-2696 (electronic)
SxI - Springer for Innovation / SxI - Springer per l'Innovazione
ISBN 978-3-031-97977-4 ISBN 978-3-031-97978-1 (eBook)
https://doi.org/10.1007/978-3-031-97978-1

Translation from the Italian language edition: "Statistica per gli indici compositi" by Matteo Mazziotta and Adriano Pareto, © Authors 2024. Published by Giappichelli. All Rights Reserved.

Cover illustration: Measuring the well-being of countries

This Springer imprint is published by the registered company Springer Nature Switzerland AG
The registered company address is: Gewerbestrasse 11, 6330 Cham, Switzerland

Preface

That September morning, the heat showed no signs of surrendering to autumn, quite the contrary. Only the light drizzle could give the impression that summer was over; in reality the humidity was at its highest, and people were perhaps suffering more than in midsummer. Classroom five was on the fourth floor. The large windows, almost like stained glass windows with decidedly unfashionable frames, let in a white, rather grey light that made the classroom seem older than it actually was. The large slate blackboard that took up almost the entire wall had been cleaned so well there were no traces of chalk, not even on the four corners or the wooden slats that surrounded it: a glossy black was just waiting to be traced with white.

The full professor of statistics was jokingly referred to as a 'fool', partly to emphasize his lack of inclination to respect classical teaching methods and partly to joke with his academic role. Some said that his teaching was always different from one year to the next, although this characteristic can be quite realistic even for teachers who are considered less 'fool' than him. In truth, they were referring to his teaching style, a kind of literary genre, as if he taught statistics in prose one year and poetry the next. Students in previous years spoke of the professor's obsession with the concept of variability, which was viewed from an anthropological, biological or physiological rather than a mathematical perspective. Any other teacher would have spoken of variability as a peculiarity of the characteristics found in the statistical units of a collective, in that they have values or attributes that are totally or partially different. This would have required students to know the following terms: characteristic, statistical unit, collective, value and attribute. Too many technical details, at least for the lessons at the beginning of the course. According to former students, the professor liked to associate variability with the concept of diversity, often using the example of genetics whereby one speaks of a characteristic of ecosystems or a pool of genes commonly considered advantageous for survival, that is, describes the existence of many different versions of the same organism. And it is this last aspect that he focused on: the organism is the same and shows up in different versions. Even if one was to study animals of the same species, race, sex, age and physical/biological characteristics, it would be impossible to obtain identical laboratory measurements between the different subjects. Some years, former students recount, this lesson could veer almost naturally towards biodiversity or biological diversity, when the professor liked to quote from memory article two of the 1992 Rio de Janeiro Convention: "*[Biological diversity is] the variability among living organisms from all sources including , inter alia, terrestrial, marine and other aquatic ecosystems and the ecological complexes of which they are part; this includes diversity within species, between species and of ecosystems*". In practice, biodiversity is the variety and richness of species in a given unit of study. In short, he argued that variability is essential to the life of species. Simply put, we exist because we are different. This belief was certainly not his own idea, but was clearly stated in "The Origin of Species" by Charles Darwin, who wrote: "*Where many species of a genus have been formed through variation, circumstances have been favourable for variation; and hence we might expect that the circumstances would generally still be favourable to variation*". What more can be added to the importance of studying, analysing, measuring, appreciating and, without exaggeration, loving variability?

When he entered the classroom that morning, the students were already in their seats, in an almost surreal silence, as if they had nothing to say to each other about statistics, the latest movie or the dreary weather. He climbed the step to the lectern and said with the satisfied smile of someone who knows he is about to amaze:

"Do you know Ashby's law?".

He gave the students a few seconds to answer, knowing that the likelihood of anyone knowing this cognitive model was slim.

"Well? Has anyone ever heard of it?".

You could see innocence and a fair amount of curiosity on the students' faces. They imagined with satisfaction that they would soon be hearing a lesson in a completely 'fool' style. And so the professor satisfied their curiosity.

"It is the law that relates the complexity (meaning the variety, the diversity) of a system to the complexity (always meaning the variety, the diversity) of those who are supposed to manage that system".

A few seconds of silence followed, during which the students tried to figure out if they really understood this story about complexity, while the professor scanned their comically confused faces. Then he spoke again.

"All in good time. Today I am going to try to write a play for you and with you, a short piece; I have never done this before, it is an experiment. The protagonists are the arithmetic mean, which we will call AM for the sake of simplicity, and the variance, which will be VAR. The two do not particularly like each other because each thinks she is more important than the other and, above all, that she can do without her. Are you ready? Let us go on stage.

AM: I am the queen of statistics! Without me, mankind could calculate practically nothing. The synthesis that I express is a fundamental element for a multitude of sciences: social sciences, economics, medical sciences... I have almost stopped counting how many there are. Starting from a set of even very many numbers, I can provide a single value that represents them all.

VAR: First of all, let me remind you that statistics is the science that deals with variability. In the sciences you mentioned, it is extremely rare that there is no variability within a phenomenon being studied. My job is precisely to explain the differences, to measure them in order to understand how big they are. Even if I were zero, that would be very important information.

AM: What you do not want to admit, because it is convenient for you, is that you are a number that no one fully understands. A positive real number with no unit of measurement: 5, 23, 1,798 of what? Years? Trees? Dollars? To make sense of it, you have to calculate the square root. And then, let us be honest, without me, who I am in you, you would not exist at all. I quote your definition from memory: with reference to a set of values, you are the average of the squares of the deviations of the individual values from their arithmetic mean. I operate twice within you: you must first calculate me on the series of values and then on the squares of the deviations. Try to do the same with the unfinished ones of the median and the mode and see if you can keep all the properties that you enjoy and for which you have only me to thank.

VAR: I do not understand why discussions with you always end like this. Your 'democratic' sense is a false advantage. Let me explain. The fact that all values contribute to your calculation has a flaw that you do not want to hear: You are sensitive to anomalous values. All it takes is for one number to be out of whack, or worse, wrong, too big or too small, and you let it affect you without fighting it. Everyone knows that the median is more robust than you because it is a positional mean and does not take into account the extreme values; and then it can provide a more accurate representation of the 'central value' for data sets with asymmetric distributions. I do not even want to talk about the mode: too many case studies, I do not like them either.

AM: I cannot believe you brought up the democracy thing. Well, I am convinced that using all modes of distribution in my calculation helps to find a value for the central tendency that better corresponds to reality, is fairer and more democratic. And what if one or more values were not wrong? What if they were actually caused by a shock? The median would ignore them. I do not. And just to be clear, you only have some use for random variables, confidence intervals and hypothesis testing, but you do not really make sense in descriptive statistics, you just go through to get to the standard deviation, which is at least expressed in a unit of measurement like I do.

VAR: So let us talk about the fact that you are compensatory, unfortunately linear, and that because of you the story of the two chickens of Trilussa has gone around the world and discredited the whole category of statistical instruments. Even those who are not at fault have to pay the price. Have you ever wondered why Trilussa never wrote a poem about the geometric mean? Because it has an inherent penalty that increases with variability and makes it smaller and smaller than you: it is not like a linear function. It is not like a chicken for a chicken: with the geometric mean the result would be zero! Is it possible that you cannot understand that the information you provide is incomplete and perhaps even misleading if you do not have a measure of variability to draw reasonable conclusions.

Curtain.

So, my friends, perhaps we now have a little more clarity about the fact that mean and variance, or any measure of variability, must coexist, that one does not exist without the other and that the only way to obtain complete information about the population of interest that we want to analyse and from which we want to draw insights about reality is to consider them together. A value of central tendency, whatever it may be, is never completely autonomous in describing a phenomenon that interests us. Similarly, it does not make sense to consider one measure of variability alone, whatever it may be. The two instruments must coexist in order to 'speak', and only in this way can we explain the full complexity of a phenomenon. What then do you say to Ashby's law that I mentioned earlier? Do we perhaps need complex tools to master complexity? And to measure it, what do we need? Think of a complex, multidimensional statistical phenomenon that is difficult to define and measure: quality of life, fragility, educational poverty. We need statistical tools that coexist, perhaps even in the same formula, that can simplify this complexity and make it easier to read. And what if the mean and a variance function could help us with this simpler interpretation of complex reality? I would like to recommend a book for you to read".

Rome, Italy

Matteo Mazziotta
Adriano Pareto

Competing Interests The authors have no competing interests to declare that are relevant to the content of this manuscript.

Contents

1 Definition of the Problem

It is generally recognized, not only in science, that the progress of a community, the development of an individual's abilities and his or her position in a complex society cannot be represented and measured exclusively in purely economic terms. Many studies conclude that economic progress does not necessarily go hand in hand with social progress, let alone quality of life or well-being. In fact, the effects of economic growth are often problems for the community that are understood as real social costs: pollution, concentration in large cities leading to the emergence of peripheral areas at risk of marginalization, alienation from work due to technological progress, the insecurity of certain social structures leading to frustration and discomfort. Recent studies by economists, sociologists and statisticians, as well as the political debate, seem to focus on the goal of linking economic growth to parameters that are difficult to restore once they have been lost, such as the balance of environmental systems, the role of man in the community, the maintenance of the basic rules of civilization. This awareness help to increase the demand (and also the supply) of information of a social nature, which also enables policy makers to understand complex realities scientifically in order to make decisions about real situations. For this reason, scientific societies, journals and publications dealing with social statistics on inequality, poverty, quality of life and well-being, both from an objective and subjective point of view, are growing by leaps and bounds, also at international level. These studies are also evolving rapidly thanks to three fundamental aspects of recent times: (1) technological progress allows us to equip ourselves with very powerful computers with great storage, management and processing capacity; (2) the spread of *Big Data* and widely accessible administrative sources, both in the form of open data and official statistics, allows us to study social phenomena down to the smallest territorial detail, radically changing the classic paradigm of data collection, which is essentially based on direct

M. Mazziotta and A. Pareto, *Statistics for Composite Indicators*, SxI - Springer for Innovation / SxI - Springer per l'Innovazione 18,
https://doi.org/10.1007/978-3-031-97978-1_1

surveys (sample and census); (3) the development of software for processing statistical data, both paid and open source, greatly simplifies the application of even complex data analysis techniques and methods.

The problem can then be simplified as 'management of complexity', as it is necessary to consider several dimensions in space and time simultaneously. Technological progress and modern times allow us to have data and IT tools that are capable of handling complicated tasks that can be solved in a limited time frame. Therefore, the management of complexity must be understood above all as a theoretical problem whose solution is presented in this text through the synthesis, or rather, the construction of composite indicators. Put simply, it is about the application of statistical tools for dimensionality reduction in order to make a complex phenomenon, which is otherwise not meaningful or even measurable, more comprehensible to the human cognitive faculty, because one of the goals of statistics is to make comprehensible that which appears undifferentiated in the mass of information. Or better still, to grasp and appreciate the variability of reality, in which everything appears mixed up or, even worse, the same. It is essentially about finding a unidimensional measure for something that is multidimensional. From an operational point of view, the construction of composite indicators allows each of n statistical units to be expressed with a single value, which simplifies the interpretation of complex phenomena otherwise expressed by m dimensions.

As already mentioned and will be repeated later in the text, composite indicators have taken a significant place in the applications of social, economic and environmental sciences in recent years, as they have been developed to measure phenomena such as quality of life, well-being, development, inequality, poverty, i.e. all those complex phenomena that can be expressed with a variety of quantitative indicators derived from reality.

1.1 Scales of Measurement

Before dealing with measurement scales, we should define what a concept is. It is a thought, as it is conceived by the mind, or better still an idea; it is a notion that represents a given reality resulting from the synopsis of the various components of a given object that the mind wishes to measure as a whole.

A concept can be explicitly observable (weight, height, income, gender, educational attainment, etc.) or not explicitly observable (quality of life, social progress, educational poverty, infrastructural facilities of an area, etc.). If the concept is represented by a single dimension in which the different 'modalities' or categories (qualitative or quantitative) are made explicit, it is called unidimensional. If the concept is represented by several dimensions, it is called multidimensional.

If the categories are quantitative, they are expressed with real numbers. On the other hand, if the categories are qualitative, they are expressed in words. Please note that it is also possible to assign numerical values to qualitative categories, but in this case they are labels or codes that cannot be used to perform arithmetic operations (Delvecchio 1995).

High-quality statistics are always based on data derived from observations or measurements, from samples, censuses or administrative sources, organized and summarized in a few simple pieces of information. This so-called empirical information must be distinguished from the theoretical information resulting from a possible statistical model that interprets the observed data.

To measure the intensity with which the categories of a concept are manifested, four types of scales are used, characterized by different properties and arithmetic operations (Stevens 1946). The measurement scales are:

- Nominal scale;
- Ordinal scale;
- Interval scale;
- Ratio scale.

The *nominal scale* is the most primitive form of measurement. It is used when it is possible to divide the measurements into distinct, mutually exclusive and exhaustive categories. Nominal scales consist of qualitative categories (labels, codes or symbols) that do not require a predetermined order. Only one relationship is defined in the nominal scale, namely that of equivalence: measurements of the same class are equivalent to each other, while all measurements assigned to distinct classes differ from each other. Given three measures a, b, and c, the nominal scale has the following properties:

1. $a = a$ (reflexive);
2. $a = b \Rightarrow b = a$ (symmetrical);
3. $a = b, b = c \Rightarrow a = c$ (transitive).

The only arithmetic operation permitted for nominal scales is counting the number of elements within each class. Examples of nominal scales are "Gender" (male, female) and "Marital status" (single, married, widowed, divorced).

The *ordinal scale* is a measurement that contains a quantity of information that is immediately greater than the nominal scale; it is based on logically consecutive categories, regardless of whether they are ascending or descending. However, it has the fundamental limitation that it does not allow quantification of the differences in intensity between the categories. In addition to the equivalence relation, the ordinal scale defines the order relation: given two measures a and b that belong to different classes, it can be determined whether $a < b$, which means "a precedes b" or $a > b$, which means "a follows b" (this also applies if the direction of sorting is reversed). The ordinal scale has the property of trichotomy, i.e. given two measures a and b, we always have:

$$a = b, \text{ or } a < b, \text{ or } a > b$$

The set of ordered classes is a 'ranking', and the ordinal number assigned to the units based on their class membership is called the 'rank'. The ranks do not directly correspond to the set of measured property, but only represent the ordering relationship. Therefore, the only operation allowed in ordinal scales is counting the frequencies of the individual classes, as in the nominal scale.

"Education" is a typical ordinal scale, as there is a logical order between the categories: (1) primary or secondary school, (2) Bachelor's or equivalent, (3) Master's or equivalent, (4) Doctorate or equivalent.

The *interval scale* links cardinal numbers to the categories and allows us to assess the width of the interval between each pair of values. In the interval scale, the unit of measurement and the point of origin are arbitrary. The unit of measurement can be changed by multiplying all values of the scale by a positive constant, and the origin can be changed by adding a constant to all values. The values of an interval scale can therefore be transformed by an equation of the form:

$$x' = \alpha x + \beta \text{ with } \alpha > 0$$

where the parameter α specifies the selected unit of measurement and the parameter β defines the origin. This ensures that the equality of the measurement differences is invariant in relation to the reference system used, i.e.:

$$(x_i - x_j) = (x_k - x_l) \Leftrightarrow (x_i' - x_j') = (x_k' - x_l') \quad \text{for each } i, j, k, l$$

It follows that with interval scales it is possible to assign the same meaning to the same measurement differences. The interval scale allows arithmetic operations based on the differences between the values associated with different points on the scale, operations that cannot be performed in the case of ordinal and nominal scales. Classic examples of interval scales are "Temperature" (measured in degrees Celsius or Fahrenheit) and "Year of birth".

The *ratio scale* is an interval scale in which the position of the zero point is not arbitrary but 'objective' and corresponds to the element with the intensity zero in comparison to the measured property. While the zero point of the interval scale is arbitrary and represents a 'relative zero', the zero point of the ratio scale is clearly defined and represents the 'absolute zero'. The unit of measurement may change, but regardless of the unit chosen, the zero point always means zero intensity. The values of a ratio scale can therefore only be transformed by the equation:

$$x' = \alpha x \text{ with } \alpha > 0$$

where the parameter α specifies the selected unit of measurement. This ensures that the ratio of the measurements in relation to the reference system used is invariant, i.e.:

$$\frac{x_i}{x_j} = \frac{x_i'}{x_j'} \quad \text{for each } i, j$$

It follows that with ratio scales it is possible to attribute a meaning of 'proportionality' to the ratio between measurements. Arithmetic operations are therefore

Table 1.1 Scales of measurement

Measurement scale	Property	Valid operations
Nominal	Categories are not ordered	$a = b$; $a \neq b$
Ordinal	Categories are ordered	$a < b$; $a = b$; $a > b$
Interval	Categories are numerical	Sum (+); difference (−)
Ratio	Categories are numerical and there is a natural zero	Sum (+); difference (−); Ratio (:); product (×)

not only possible on the differences between the values of the scale (as with the interval scale), but also on the values of the scale itself. Typical ratio scales are "Weight", "Height", "Age", "Distance", "Speed" and "Income".

Table 1.1 provides an overview of the most important properties of the measurement scales. It is clear that there is a cumulative hierarchy between them, i.e. each scale allows us to capture information more accurately than the previous one. It follows that the higher the scale, the more precise the statistical methods used. For this reason, when measuring concepts, the highest type of scale that the phenomenon to be observed allows must be used.

A final consideration in relation to measurement scales concerns the number of categories into which they can be divided. For qualitative scales (nominal and ordinal scales), the categories necessarily form a finite set, whereas for quantitative scales (interval and ratio scales) their number can be finite or infinite, i.e. they can be represented by countable numbers (discrete variable) or have infinite values (continuous variable).

1.2 Indicators and Indices

Over time, the literature has produced many definitions where it is difficult to agree on a common assumption about the difference between indicator and index, to which the notion of variable should also be added for the sake of clarity. It is certainly not the task of this book to find a solution to a problem that has been discussed by so many illustrious representatives over the years, but the aim is to find common sense definitions that can guide us in our reading, especially in the more analytical parts, so that there are no misunderstandings.

The term *variable* is derived from the late Latin *variabilis* and means "that which varies, that which tends to vary". In statistics, it means a characteristic that is collected in one or more statistical units of a population or a sample as the result of a direct survey. For our time, this definition seems almost too short-sighted, as we live in the age of *data deluge*, where we are surrounded by big data or administrative sources that are used for statistical purposes and thus not only for direct survey. We can therefore say more generally that the variable is defined as a characteristic associated with a statistical unit (e.g. the age of a person or the number of hospital beds in a region) and can be measured on a nominal, ordinal, interval or ratio scale.

The term *indicator* is derived from the late Latin *indicātor* and means "that which points out, someone who indicates something or an element that signals something". Economic indicators are macroeconomic magnitudes that are considered significant for assessing the economic performance of a particular country (e.g. GDP per capita, interest rate, inflation rate, etc.). Social indicators are statistics used to measure significant variations in the state of important aspects of a society (e.g. life expectancy, mean years of schooling, crime rate, etc.). Here it becomes clear that the indicator is a fraction and is therefore composed of a numerator, which can be seen as the variable that provides the meaning, and a denominator, which is the variable that enables its comparability in space or time. Of course, both variables must be measured on an interval or ratio scale. Let us take the example that we want to compare the level of health infrastructure of two countries, Italy and Malta. If we were to look at the variable "Number of hospital beds", the comparison would not be meaningful, as Italy naturally has more beds than Malta. However, if we calculate the indicator "Number of hospital beds per 1000 inhabitants", we obtain a value normalized to the population variable and can then make a comparison between the two countries. In other words, the presence of a reference denominator (the population figure) makes it possible to transform a variable (the number of beds) into an indicator, i.e. a relative measure that is comparable in space and time.

Finally, the Latin term *index* means "anything that is used to inform". By index we mean the synthetic expression of the dimensions of a particular phenomenon in space and/or time or the relationship between several phenomena considered simultaneously. The index could therefore be defined as a complex structure in which the dimensions, also (but not necessarily) with different units of measurement, are summarized by mathematical operations. For example, the Gini Index is a measure of the degree of inequality in the distribution of a transferable variable, such as income, among the n units of a community. The various formulas in the literature for its calculation offer solutions ranging from the use of the relative mean absolute difference to the areas of the Lorenz curve. Another example is the Body Mass Index (BMI) of a person, which is calculated as the ratio between the weight and the square of the height (variables that have different units of measurement). A further example is the Human Development Index (HDI), which is calculated as the geometric mean of the normalized indices of the three dimensions of health, knowledge and standard of living. These three examples show us that the index is different from the indicator, as it is a more complex measurement that involves operations with variables or indicators.

In the literature we generally find the term *individual indicators* to express the basic indicators, i.e. the indicators as we have described them in the previous paragraph, and *composite indicators* to define the indices, i.e. the combinations of individual indicators.

In this book, we use the term 'indicator' or 'individual indicator' to refer to a basic indicator that represents a single component of the complex phenomenon to be measured, and the term 'composite indicator' or better 'composite index' to indicate a composite measure of the phenomenon itself.

1.3 Types of Indicators

The classification of indicators is another much-discussed topic on which there is an extensive literature, especially in the social domain. Since composite indices apply to different types of statistical indicators (including economic and environmental indicators), we present some general classifications (Aureli Cutillo 1996) in order to focus on the most useful indicator categories for the purposes of synthesis in the following two sections.

Statistical indicators can be categorized into several broad groups relevant to the objective of the study:

1. From the point of view of the relationship between the indicator and the complex phenomenon, we have:
   - *Constitutive indicators*, which occur when the complex phenomenon can be decomposed into a set of dimensions that constitute it and that are individually measurable by indicators.
   - *Concomitant indicators*, which occur when the complex phenomenon has a concomitant behavior with a unidimensional phenomenon; this concomitant behavior can be determined by a dependency relationship or a common cause.
2. From the point of view of purpose, we have:
   - *Descriptive indicators*, which consist of data collected to describe a complex reality; they are used primarily for explanatory and comparative purposes in space and time.
   - *Normative indicators*, which consist of standard ratios against which the empirical indices are compared and must evolve; they are mainly used for planning purposes and serve to evaluate expenditure, investments, allocations or redistributions of resources. Not all indicators of this type are required by law, they may also indicate a goal that is considered desirable.
   - *Predictive indicators*, which contain an indication of a rising or falling trend. They are usually expressed in terms of relative differences, and the more variable they are, the more useful they are for their predictive power.
3. From the point of view of the relationship between the measure and the reference period, we have:
   - *Static indicators*, which indicate the measured value or level at a specific point in time (e.g. a day, a month, a year). The data are static in the sense that they describe a phenomenon as it occurred at a given time.
   - *Dynamic indicators*, which indicate the degree or extent of change during a specific period of time (e.g. from one year to the next). The data are dynamic in the sense that they describe a variation—increasing, decreasing or constant—over time.

4. From the point of view of the information source, we have:
   - *Objective indicators*, which are based on directly observable and statistically measurable raw data relating to individually identifiable facts and social phenomena.
   - *Subjective indicators*, which are based on the perceptions of individuals and therefore represent personal assessments.

However, the proposed classifications do not exhaust all the ways of grouping indicators that can be found in the extensive literature on the subject and have also been created by independent researchers depending on specific research topics.[1]

1.3.1 Objective and Subjective Indicators

The idea that capturing subjective aspects has a high informative and analytical value is now widely accepted. Indeed, perceptions and evaluations influence the way people approach life and the opportunities it offers them. Subjective indicators are a useful complement to strictly objective indicators as they allow us to assess discrepancies between what people report and what is captured by objective indicators. Taking these indicators into account allows for a clearer and more comprehensive view, especially in terms of describing the phenomenon under study. In recent decades, public organizations have preferred quantitative, objective, indicators (hard indicators) to measure social 'concerns', mostly derived from censuses or sample surveys of institutional subjects (e.g. demographic rates, number of inhabitants per doctor, average number of members per household). However, the usefulness of subjective-qualitative or perceptual indicators (soft indicators) derived from psychological approaches has been supported for some time. They allow to measure individual experiences, identify people's perceptions of the social world and obtain information about their attitudes, needs, desires, expectations, perceived well-being or discomfort, satisfaction or dissatisfaction with aspects of domestic life and social relationships, social issues, personal and collective security, employment prospects, discrimination phenomena of all kinds, public spending targets and the behavior of government officials.

The use of these two types of indicators in the construction of composite indices is very different and deserves the greatest attention of the researcher. As a rule, objective indicators are used more frequently because they enjoy a high level of credibility and are mainly constructed by institutional subjects according to standardized procedures defined by international organizations. They are derived from sample surveys, censuses or administrative sources treated with techniques tested and approved by the international scientific community. Essentially, they are created from statistical data on social, economic or environmental events that are directly verifiable because they are real. Subjective indicators arise from the collection and processing of opinions, judgments, beliefs, feelings, attitudes, desires,

[1] See, for example, Uwizeyimana (2021).

perceptions, predictions and motivations of individuals who are part of the population being studied. What are the problems behind the subjective indicators? It seems particularly difficult to verify the desirable properties of an indicator such as reliability and validity. Indeed, it should bear in mind that subjective data, which vary from respondent to respondent and even for the same respondent depending on the historical moment in which the question is asked, do not allow, on the one hand, a verification of uniqueness that would make it possible to confirm their reliability and, on the other hand, are characterized by a complicated extension to the population. In principle, it always seems difficult to carry out a direct verification of their validity. Over the years, there have been numerous attempts to classify subjective data using various scales, which will not be discussed in detail here. However, the difficulty of using these types of indicators in the classical paradigm of constructing composite indices remains, as they are usually based on ordinal scales.

1.3.2 Input, Output and Outcome Indicators

As already described, we speak of individual indicators when they refer to individual phenomena, i.e. to the directly quantifiable dimensions into which a complex phenomenon is broken down, while we speak of composite indices when they are obtained by aggregating and weighting the individual indicators that form the basis for measuring complex phenomena.

Indicators are therefore tools that can show (measure) the evolution of a phenomenon considered representative for the analysis and that serve to monitor or evaluate the success or adequacy of the actions carried out. An indicator is a measure that is generally expressed in quantitative form and consists of several variables that can summarize the evolution of the phenomenon to which it refers. The indicator is not the phenomenon itself, but represents the behavior of the phenomenon to be observed and evaluated. To give an illustrative example: "The indicator is the finger that stretches to the sky… but the phenomenon is the star!".

In the previous sections, some classifications of indicators have been proposed. This proposal is now fundamental to the theory of constructing composite indices: input, output and outcome indicators. Not all of these indicators are suitable as individual indicators for the calculation of a composite index. Let us see in detail why.

The input indicator refers to all elements that make up the resources and the baseline situation (including needs and risk factors) that can be understood as the basis for planning and program design. In general, inputs are the resources (usually economic and financial) that are put into a process to achieve a result in a given time, e.g. the expenditure required to achieve a goal. The output is the most direct result of the expenditure program. The outcomes express the impact that the spending program has on the community and the environment. Output and outcome indicators are characterized by strong interdependencies, and whoever uses them must still take into account the influence of exogenous factors.

In the literature, a distinction is often made between product indicators (output) and impact or result indicators (outcome). Objectives and indicators can refer to the products (output) or to the resulting effects (outcome) or to a suitable combination of both. The product is that which is directly created by the program (e.g. a service for a specific user). In a broader sense, the outcome can be seen as a consequence of the product. Of course, there must be a plausible and close relationship between this 'result' and the 'product' as a consequence of which it is considered. Otherwise, it is the result of external factors and the researcher may come to false conclusions that do not correspond to reality.

We will explain the difference between the three types of individual indicators using an example. Let us imagine that we want to increase the employment rate in a certain geographical area as part of a socioeconomic development program. The goal is to provide a specific target group with the necessary skills to find a job by participating in vocational training courses. The input indicator represents the expenditure made to organize the refresher course (classroom, teaching material, teachers, etc.), the product (output indicator) is represented by the people who successfully completed the course, while the effect (the result or outcome indicator) is represented by the people who found a job this way and increase the employment rate. If we want to measure a complex phenomenon with a composite index, we must be extremely careful when inserting an input indicator into the original matrix, as it is unlikely to interpret its role as a component of the phenomenon well. If vocational training shows a sharp decline in the number of participants or simply does not help learners to find a job, this means that the expenditure made, and therefore the investment, has not produced the desired results. There is therefore a risk of inserting an indicator into the matrix whose information content is completely useless or, worse still, incorrect, as the objective for which the expenditure was made has not been achieved. If we insert the output indicator in the original matrix, in our example the learners who have completed the course with profit, we could say that this indicator serves perfectly to compose the complex phenomenon by calculating the composite index, as it would be a significant product of the expenditure made; the participants have acquired a wealth of knowledge that they can use in their job search. The ideal would be to look at the outcome indicator, i.e. the end result of the expenditure made, in our case the increase or decrease in the employment rate.

1.4 Why the Synthesis?

In recent years, the debate on the measurement of multidimensional phenomena has aroused renewed interest in the scientific community of many industrialized countries. It is generally recognized that many socioeconomic phenomena cannot be measured by a single descriptive indicator, but must be represented by a variety of aspects or dimensions. In order to measure phenomena such as development, progress, poverty, social inequality, prosperity, quality of life, infrastructure provision, etc., different dimensions must be 'combined' together as components of the phenomenon. Indeed, the complex and multidimensional nature of these phenomena requires the definition of objectives, the achievement of which can be observed and measured by individual indicators. The mathematical combination (or aggregation, as it is defined) of a set of indicators representing the different dimensions of a phenomenon to be measured can be achieved by applying statistical methods that lead to results known in the literature as composite indices (Aureli Cutillo 1996; Booysen 2002; Saisana and Tarantola 2002; Salzman 2003; OECD 2008; Terzi et al. 2021).

The construction of a composite index is a delicate task and involves many pitfalls, such as the availability of data, the selection of individual indicators and their treatment in order to compare (normalization) and combine (weighting and aggregation) them. Nevertheless, composite indices are widely used by various international organizations to measure economic, environmental and social phenomena and are therefore a very important and evolving tool (OECD 2008).

Many scientists reject the use of composite indices that lead to the determination of a single value for each statistical reference unit (e.g. a geographical area) and prefer the so-called 'dashboard' (as in the case of monitoring the operating condition of a vehicle: oil level, gasoline, water temperature, etc.). In the case of the dashboard, different manifestations of the phenomenon can be identified, all of which are relevant but not further aggregated. From a statistical point of view this makes sense, but from a political and media point of view it is a serious limitation. Ease of dissemination in the media and immediate understanding by the user are certainly the strengths of a single index to measure a complex phenomenon. Obviously, both approaches have strengths and weaknesses. The dashboard handles complexity without the use of composite measures and certainly has drawbacks in terms of communication. If we were to ask ourselves the following question when measuring the complex phenomenon of well-being after looking at a dashboard with individual indicators, we would probably not have a clear answer: "Is well-being increasing or decreasing?". The composite index also copes with complexity, but reduces the dimensions in the space, which inevitably leads to a loss of information. However, it offers a 'tolerated simplification' of reality and allows even an uninformed audience to understand the message by easily recognizing an increase or decrease in the phenomenon. If we still want to take well-being into account, we should be able to answer another question that goes beyond theoretical and methodological aspects: "Is it possible to measure well-being with a formula?". The answer is probably yes, if we adheres strictly to a work paradigm (see Chap. 3). There are numerous attempts in the literature to measure well-being that do not adhere to a correct work paradigm and therefore come to unreliable and questionable conclusions. This very aspect that was and is the reason for the failure of many alternative measures to GDP.

The above debate has convinced several scholars that the macroeconomic measure par excellence, GDP, is not capable of representing the well-being or progress of a society, let alone expressing the quality of life of a geographical area or a community of citizens. This debate has generated a considerable literature worldwide with more than a hundred alternative indices published by governmental organizations (and not only), academia and the business press, but despite this, the popularity of GDP has not suffered in the slightest. Indeed, GDP is based on very solid theories, while many alternative indices are amibiguos in the very definition of the phenomenon and in many cases, since they are not based on a common socioeconomic theory, take into account dozens of indicators, so that all possible aspects (and therefore none) are considered. The publication in September 2009 of the report of the Commission on "The Measurement of Economic Performance and Social Progress", also known as the Stiglitz Commission, set up by former French President Nicolas Sarkozy, was fundamental to the development of numerous studies on "beyond GDP" scenarios. One of the Commission's objectives was to demonstrate the limits of GDP as the sole indicator of economic performance and social progress. It also aimed to investigate what other information could be asked in the current questionnaires to create more relevant indicators of social progress, to assess the feasibility of alternative measurement tools for the phenomenon and to discuss how to present the information statistics appropriately (Giovannini and Rondinella 2012). In fact, even before the Stiglitz Commission, several attempts to measure phenomena 'close' to well-being (progress, quality of life, happiness, etc.) were published by researchers and prestigious institutions. These experiments can be divided into four groups, some of which are presented below.

The first group proposes an adjustment of GDP and includes: Measure of Economic Well-being (MEW) (Nordaus and Tobin 1972), Index of Sustainable Economic Well-being (ISEW) (Daly and Cobb 1989) and Genuine Progress Indicator (GPI). The second group of studies deals with aspects such as social and environmental activities or directly with the degree of (perceived) satisfaction of individuals and includes: Happy Life Expectancy (HLE) (Veenhoven 1996) and Happy Planet Index (HPI). A third group concerns a composite index that includes GDP, such as Human Development Index (HDI) (UNDP 1990, 2001, 2010), Better Life Index (BLI) (OECD 2011) and Gross National Happiness (GNH). Finally, the fourth approach states that it is better to measure different dimensions with a set of indicators (dashboard) than to obtain a single composite measure (Rinaldi and Zelli 2014), for example the Millennium Development Goals (MDGs) of the United Nations Development Program (UNDP). For a detailed overview of composite indices, see Bandura (2008).

In Italy, the first report on Equitable and Sustainable Well-being (BES) was published in March 2013, the result of a collaboration between the Italian National Institute of Statistics (Istat) and CNEL (National Council for Economy and Labour). It is a dashboard with around 130 individual indicators collected in 12 areas. The third BES report, published in December 2015, is the first to present a composite index for each of well-being outcome areas (Istat 2015). For further details, see Chap. 5.

1.5 Properties and Characteristics of a Composite Index

The construction of a composite index is an excellent way of capturing non-unidimensional phenomena, although a work paradigm must be strictly adhered to. It is a difficult task with many decisions to be made and alternatives to be evaluated, which may even strongly influence the quality and accuracy of the results. Among the main problems of this approach are the choice of the theoretical framework, the availability of data (spatial and temporal), the selection of the most representative indicators and their treatment in order to compare and aggregate them (Fanchette 1974).

The work paradigm is mainly based on the following steps (Salzman 2003; OECD 2008; Mazziotta and Pareto 2017):

1. Definition of the phenomenon. The conceptual definition should provide a clear idea of what the composite index is intended to measure. It should also refer to a theoretical framework that has been 'thought of' by one or more groups of experts on the topic in question.
2. Selection of the individual indicators. Ideally, indicators should be selected on the basis of their relevance, analytical robustness, timeliness, accessibility, etc. (Maggino 2014). It is important to keep in mind that many socioeconomic phenomena such as well-being are based on a formative model, which corresponds to the assumption that the latent factor (in this case well-being) depends on the individual indicators that 'explain' it, and not vice versa (Diamantopoulos et al. 2008).
3. Exploratory analysis. It allows the characteristics of the individual indicators to be examined and can therefore provide useful information on the methods to be used for normalization and aggregation.
4. Normalization of the individual indicators. This step aims to make the indicators comparable with each other and to define their 'polarity'. Normalization is necessary before any aggregation, as the indicators are very often expressed in different units of measurement. Moreover, it is necessary to normalize the individual indicators in such a way that an increase in the normalized indicators corresponds to an increase in the composite index.
5. Aggregation of the normalized indicators. This is the combination (mathematical function) of all components to obtain the composite index. Of course, different aggregation methods are possible, and the choice of these methods must depend on the type of indicators within the measurement model under consideration.
6. Validation of the composite index. The aim here is to assess the robustness of the composite index in terms of its ability to produce stable measures.
7. Presentation of the results. The presentation of a composite index is not a trivial matter. A good presentation in tabular or graphical form must help to effectively convey the message it contains and improve the interpretation of the measured phenomenon.

It is important to emphasize that the theoretical part is not separate from the statistical-methodological part; in fact, the definition of the phenomenon and the selection of individual indicators are not independent of the choice of normalization and aggregation methods. Unfortunately, many studies in the literature do not take this aspect into account and consider, for example, the use of principal component analysis or, even worse, factor analysis as a synthesis method in formative models (see Sects. 3.5.3 and 3.8.1).

There is no universal method for the construction of composite indices. Their construction depends on many factors, including the area of application, the target group to which they refer, the objectives of the research, the dissemination in the media and much more. In any case, the advantages of composite indices are obvious, as they provide a unidimensional measure of a complex phenomenon and therefore allow for easier interpretation than a set of individual indicators, which simplifies data analysis.

A basic rule we should always keep in mind is "garbage in, garbage out", i.e. if the original matrix contains obvious errors, then the composite index will also contain errors. If a phenomenon is poorly defined, it will certainly be poorly measured, but the reverse is certainly not true. If the phenomenon is well defined and the original matrix consists of individual indicators of good quality, the composite index is not always valid. It also depends on the statistical methodology used, which must be well compatible with the theoretical framework on which the phenomenon to be measured is based.

As already mentioned, there is no universal method for constructing composite indices. The best composite index is the one that takes into account the objectives desired by the researcher and/or the client. The work paradigm requires that we ask ourselves some questions before we begin. The answers to these questions will influence the path we need to take to get the best possible solution for the composite index. All answers can influence both the selection of individual indicators and the methodology for their normalization and aggregation. In particular:

- Do we need spatial comparisons? If so, the selected individual indicators must be available for the desired level of disaggregation of the area, which may or may not affect the use of some measures.
- Do we need temporal comparisons? If so, the normalized individual indicators must be available for the desired time series, and this condition can only be met with some normalization methods.
- Are the individual indicators not interchangeable? Or is a compensation between them permissible? When measuring socioeconomic phenomena, a formative model is usually used, so that compensation is absolutely undesirable. If the individual indicators are not interchangeable, this must be taken into account when choosing the aggregation method. In this case, the arithmetic mean or linear models are generally not correct. For example, the Human Development Index (HDI) is characterized by non-substitutable individual indicators that cannot be fully compensated for by the aggregation method (the geometric mean).

- For which audience is the analysis intended? The client and recipient of the composite index should influence the choice of method for aggregating the individual indicators. Simple calculation, immediate application and easy interpretation of the results are essential requirements if the study is aimed at a broad audience unaccustomed to technical details; the reader must immediately understand both the methodology used and the meaning of the results obtained. If the study is aimed at an academic audience, the methodology may well be more complex and the results may have different 'nuances of interpretation'. In any case, the transparency of the method and the calculation must be maintained, otherwise the composite index is a fraud.
- Is the method robust? The first rule for creating a good composite index is to adhere to the objectives of the study. But other rules must also be observed. The index must be robust, i.e. it must be sensitive to changes in the individual indicators and must not be overly influenced by anomalous values. The method must be stable (but not too stable) against variations in the input matrix. It is very important to select the most reliable method by uncertainty analysis (influence analysis or other similar techniques).

The answers to these questions should guide research towards the most effective method to reduce the multidimensionality of the phenomenon. It is not possible to ignore any of these questions, because the risk of altering reality is very high. Particular attention must be paid to finding the most appropriate method based on the following factors (Mazziotta and Pareto 2013): type of model (reflective/formative), type of indicators (substitutable/non-substitutable), type of aggregation (simple/complex), type of comparisons to be made (absolute/relative), type of weights of the indicators (objective/subjective), as described in Sect. 3.8.1.

1.6 Advantages and Disadvantages of a Composite Index

Every day, our behaviors can be seen as composite indices. When we have to make a decision, we evaluate the circumstances, the hypotheses and the appropriateness of the different actions. The final decision is the result of the simultaneous weighing of many factors and their weighting according to our own view of things. Without realizing it, composite indices are part of us and are part of our daily way of acting.

Reducing the dimensions of space to a single one is certainly a daring and complex endeavor that involves many pitfalls, such as the loss of information or even just the simplistic interpretation of the final results, but it is necessary so that we can more easily grasp a multidimensional reality. Managing complexity is the real challenge for the researcher of any scientific discipline, which does not necessarily mean resorting to the calculation of composite indices. Even if we wanted to represent reality through a set of individual indicators, the so-called dashboard, without resorting to synthesis, we would still have to deal with complexity while taking into account a set of elements that are also very diverse.

The increase in publications, studies and research on the measurement of complex phenomena in recent years has drawn attention to various methodological tools and the appropriateness of their use in both academic and policy forums.

The advantages of composite indices are considerable and are summarized below (OECD 2008):

- Can summarize complex realities to assist decision makers.
- Are easier to interpret than a battery of individual indicators.
- Allow to assess the progress of statistical units (geographical areas or others) over time.
- Reduce the visible dimension of a set of individual indicators without losing fundamental information.
- Put questions about a country's performance and progress at the center of the politcy debate.
- Facilitate communication with the public (citizens, media, etc.).
- Contribute to shaping the opinions of target groups with different backgrounds.
- Enable users to effectively compare complex dimensions.

However, the disadvantages and risks of composite indices are manifold and can even be very serious. In particular:

- Can send misleading messages if poorly constructed or interpreted.
- Can lead to simplistic political conclusions.
- May be used fraudulently (e.g. to support a particular policy decision) if the construction process is not transparent and/or lacks sound statistical or conceptual principles.
- The selection of indicators and their weighting can be the subject of heated (scientific and/or political) debate.
- May mask serious gaps in some dimensions, if the construction process is not transparent.
- May lead to inappropriate intervention measures if some hard-to-measure performance dimensions are ignored.

From the advantages and disadvantages listed, it is clear how important it is to create composite indices according to a rigorous workflow in which the decisions at each stage are well justified by the objectives of the measurement. The unidimensional measurement of a multidimensional phenomenon is a very powerful tool to read what hardly seems clear to our eyes, and therefore also very risky if it is not well constructed. This text is intended to be a manual that guides the reader through the best practices to follow in order to obtain statistically valid and robust composite indices.

References

Aureli Cutillo E (1996) Lezioni di statistica ufficiale, Parte seconda. Sintesi e graduatorie. CISU, Rome

Bandura R (2008) A survey of composite indices measuring country performance: 2008 update. A UNDP/ODS Working Paper, New York

Booysen F (2002) An overview and evaluation of composite indices of development. Soc Indic Res 59:115–151

Daly HE, Cobb JJ (1989) For the common good: redirecting the economy toward community, the environmental and a sustainable future. Beacon Press, Boston

Delvecchio F (1995) Scale di misura e indicatori sociali. Cacucci Editore, Bari

Diamantopoulos A, Riefler P, Roth KP (2008) Advancing formative measurement models. J Bus Res 61:1203–1218

Fanchette S (1974) Social indicators: problems of methodology and selection. In: Unesco, social indicators: problems of definition and of selection. Reports and papers in the social sciences, vol 30, Paris

Giovannini E, Rondinella T (2012) Measuring equitable and sustainable well-being in Italy. In: Maggino F, Nuvolati G (eds) Quality of life in Italy: research and reflections. Springer, New York, pp 27–40

Istat (2015) BES 2015. Equitable and sustainable well-being in Italy. Istituto nazionale di statistica, Rome

Maggino F (2014) Indicator development and construction. In: Michalos AC (ed) Encyclopedia of quality of life and well-being research. Springer, Dordrecht, pp 3190–3197

Mazziotta M, Pareto A (2013) Methods for constructing composite indices: one for allor all for one? Ital J Econ Demogr Stat Stud LXVII(2):67–80

Mazziotta M, Pareto A (2017) Synthesis of indicators: the composite indicators approach. In: Maggino F (ed) Complexity in society: from indicators construction to their synthesis. Springer, Cham, pp 159–191

Nordaus WD, Tobin J (1972) Is growth obsolete? National bureau of economic research, economic research: retrospect and prospect. Columbia University Press, New York, pp 509–564

OECD (2008) Handbook on constructing composite indicators. Methodology and user guide. OECD Publications, Paris

OECD (2011) Compendium of OECD well-being indicators. OECD Publications, Paris

Rinaldi A, Zelli R (2014) Misurare il benessere. Donzelli editore, Rome

Saisana M, Tarantola S (2002) State-of-the-Art report on current methodologies and practices for composite indicator development. EUR 20408 EN

Salzman J (2003) Methodological choices encountered in the construction of composite indices of economic and social well-being. Center for the Study of Living Standards, Ottawa

Stevens SS (1946) On the theory of scales of measurement. Science 103:677–680. http://www.jstor.org/stable/1671815

Terzi S, Otoiu A, Grimaccia E, Mazziotta M, Pareto A (2021) Open issues in composite indicators. A starting point and a reference on some state-of-the-art issues. Roma TrE-Press, Rome

UNDP (1990) Human development report 1990: concept and measurement of human development. Oxford University Press, Oxford

UNDP (2001) Human development report 2001. Oxford University Press, New York

UNDP (2010) Human development report 2010. Palgrave Macmillan, New York

Uwizeyimana DE (2021) A new typology of monitoring and evaluation indicators. Eur J Econ Law Soc Sci 128–154

Veenhoven R (1996) Happy life-expectancy. A comprehensive measure of quality of life in nations. Soc Indic Res 39:1–58

Theoretical Model

2

In the field of research with which this book is concerned, measurement is the quantification of any multidimensional phenomenon so that it can be more easily read and thus interpreted in order to obtain an overview of a complex reality. One of the most important steps is to establish a link between the concept (global definition and delimitation of the object of measurement) and the observable reality (data). Measurement is therefore a kind of '*redde rationem*' of abstract concepts, in which the researcher can establish a link between theory and practice on the one hand and assign a value (or score) to the phenomenon in space and time on the other. Once this is done, any abstract assumption can be measured.

It turns out that theory and measurement are inevitably linked and that a theory is incomplete without measurement and vice versa. The assumption that every poorly defined phenomenon is also poorly measured always remains valid.

Thus, we start from the assumption that it is necessary to define the concept we want to measure as reliably as possible; the researcher chooses the theory (or theoretical framework), regardless of whether it already exists or whether it has original features. The process of operationalization is then carried out, i.e. the translation of the abstract theoretical concepts into concrete, defined and observable measures. It consists in operationalizing a theoretical construct, i.e. a well-defined abstract concept, so that it can be measured and studied empirically. Suppose we want to study the concept of 'intelligence'; standardized tests such as IQ (Intelligence Quotient) or measures of scholastic/academic performance could be used for this purpose. If we try to study the concept of 'well-being', we could operationalize it using objective indicators (health, work, income, etc.) and/or subjective indicators (perception of life satisfaction, leisure time, future prospects). In summary, operationalization is of great importance in empirical research because it allows us to move from abstract concepts to concrete and observable measures that can be analysed and studied in a scientific way.

M. Mazziotta and A. Pareto, *Statistics for Composite Indicators*, SxI - Springer for Innovation / SxI - Springer per l'Innovazione 18,
https://doi.org/10.1007/978-3-031-97978-1_2

Since the indicators that constitute the phenomenon under study are achieved through the operationalization of the concept, it is good to know that it: (i) cannot be 'perfect' but 'optimal', in the sense that it better links the theoretical framework to the availability of indicators in space and time required for the research; (ii) must be made explicit and justified so that everyone can understand the fundamental transition between the abstract concept and the measurements that are supposed to represent it empirically; (iii) is necessary, but always insufficient in relation to the theoretical complexity of the concept; (iv) must not concretize the abstract, i.e. abstract concepts, categories and ideas must not be seen as concrete, and it must not be 'reified'; (v) is the objective side of the scientific research.

2.1 Latent Variable

In recent decades, the concept of latent variables has been the subject of an interesting debate, not only among statisticians, which has led to an extensive literature of both theoretical and, above all, applied nature. In the social sciences and psychometrics, for example, the concept of latent variables has been studied in depth in order to find a scientific solution for measuring (in the sense of quantifying) phenomena that cannot be directly observed. The first author to introduce the concept of latent variables was Charles Spearman in his 1904 article in the "American Journal of Psychology" defining the concept of intelligence. From here, a common definition can be summarized as follows: A variable is defined as latent if it cannot be measured directly. For example, concepts such as social exclusion, intelligence, well-being or a person's quality of life cannot be measured directly. However, much of the current research, including official statistics, is specifically interested in quantifying these concepts by starting from a multidimensional perspective and selecting a set of measurable individual indicators.

Latent variables, also known as unobserved or hidden variables, are essentially theoretical constructs that are not directly measured but derived from observed data. These variables are used to represent concepts or characteristics that are difficult to measure directly due to their complexity.

A latent variable has the following characteristics:

- It is unobservable because it cannot be observed or measured directly by a single indicator.
- It represents an underlying construct that is assumed to influence or be influenced from the observed phenomena.
- It is associated with observable indicators that are used to indirectly assess the latent construct.
- It is in many cases multidimensional, i.e. it consists of several underlying factors or dimensions.
- It can be included in statistical models to capture and account for the relationships between the observed variables.

It is possible to distinguish different types of latent variables, just as we classify classical variables: continuous, ordinal, categorical, binary.

The measurement of latent variables involves two main steps. In the first step, observable individual indicators are selected that are assumed to reliably represent the latent construct. In the second step, observable individual indicators are inserted into a statistical model in order to arrive at a representation (measurement, in the case of composite indices) of the complex phenomenon.

Latent variables are of great importance in various areas of research, e.g. in modelling complex constructs, reducing measurement error, understanding theoretical and statistical relationships, predictive modelling, psychological and educational assessment. In this text, we will mainly focus on research in the economic and social domain, where latent variables help to explain complex socio-economic phenomena such as poverty, inequality and well-being of people. There are various statistical methods for analysing latent variables, but in this text we will focus on composite indices.

Latent variables are crucial for understanding complex, multidimensional phenomena and modelling unobservable constructs, as they reveal hidden patterns in the data that are not clearly and immediately visible. As we have already mentioned and will see in the next chapters, they are used in different areas of research. Thanks to the use of techniques for constructing composite indices, it is possible to better recognize the complex relationships underlying the phenomena under investigation.

It is clear here that the research field of latent variables is constantly evolving in response to emerging trends. With the availability of large datasets, we are investigating how latent variables can be used to extract information from big data and recognize phenomena that were not visible until recently. The availability of spatial and temporal databases (also thanks to the new season of official statistics) allows the application of dynamic and longitudinal models containing latent variables to capture changes and dynamics over time. Bayesian methods are increasingly used to estimate latent variable models, providing a probabilistic framework for quantifying uncertainty. In addition, integration with *machine learning* techniques, such as *deep learning*, opens up new possibilities for modelling complex latent structures. And last but not least, artificial intelligence, which is already a scenario waiting to be discovered and is developing rapidly and continuously.

2.2 Measurement

Measurement is the process by which, in our case, a multidimensional phenomenon is quantified; the measure, i.e. the composite index, is the result of this process. Measurement makes it possible to know, describe and control a phenomenon and to make decisions to change its state.

Without going into the science of measurement, which is very old because it has always been a need of man who relates to phenomena and other people around him, Galileo Galilei stated between the sixteenth and seventeenth centuries: "*We count what is countable, measure what is measurable and what is not measurable make measurable*". It is precisely this definition by Galileo that is ideal for measuring a latent variable. Lord Kelvin stated in the second half of the nineteenth century: "*When you can measure what you are speaking about, and express it in numbers, you know something about it; but when you cannot measure it, when you cannot express it in numbers, your knowledge is of a meagre and unsatisfactory kind*". Blalock (1982) coined a definition of measurement that fits perfectly with the main theme of this text, namely transformation for data synthesis: "*By measurement, we refer to the general process through which numbers are assigned to objects in such a fashion that it is also understood just what kinds of mathematical operations can legitimately be used*". Following Blalock, we therefore define measurement as the process of linking abstract concepts to empirical indicators.

Abstract concepts do not have a one-to-one correspondence with empirical indicators and can be operationalized and measured in an almost infinite number of ways. Since they cannot be directly observed or measured, they must be represented by empirical indicators. Unlike concepts, empirical indicators are as specific, as precise and as limited as the theoretical formulations allow. In other words: Indicators are meant to approximate concepts empirically (Zeller and Carmines 1980).

Measurement is a complex process that is first developed theoretically and then put into practice. In order to measure, one must therefore develop a mental model of the phenomenon. The model itself influences the choice of the instrument to be used and the procedure for selecting the individual indicators. The type of model depends on the purpose for which the indicators are selected. For this reason, there are no better or worse models, only more or less effective models for representing the phenomenon under study. To summarize, the objectives for which a measurement is carried out are as follows:

- Monitor a phenomenon (spatially and/or temporally).
- Improve the understanding of a phenomenon that is only partially known.
- Check the quality of the measurement instrument (statistical model).

The conceptual model defines the phenomenon to be studied and the general aspects that characterize it in the phases of operationalization and measurement. In a botanical analogy, one could say that the tree represents the abstract concept, the branches are the dimensions (or *pillars*) and the leaves of each branch are the individual indicators. In constructing a composite index, we first define the phenomenon we want to measure (definition of the theoretical framework), then identify the dimensions relevant to the interpretation of the phenomenon (first level of operationalization), and finally select the individual indicators that represent each dimension (second level of operationalization) (see Sect. 3.1).

The gap between abstract theoretical concepts and measurable empirical indicators is bridged by the "auxiliary theory" (Blalock 1968), a theory related to the measurement model. A measurement model describes the relationships between a construct and its measures (indicators), and the opposite direction of causality leads

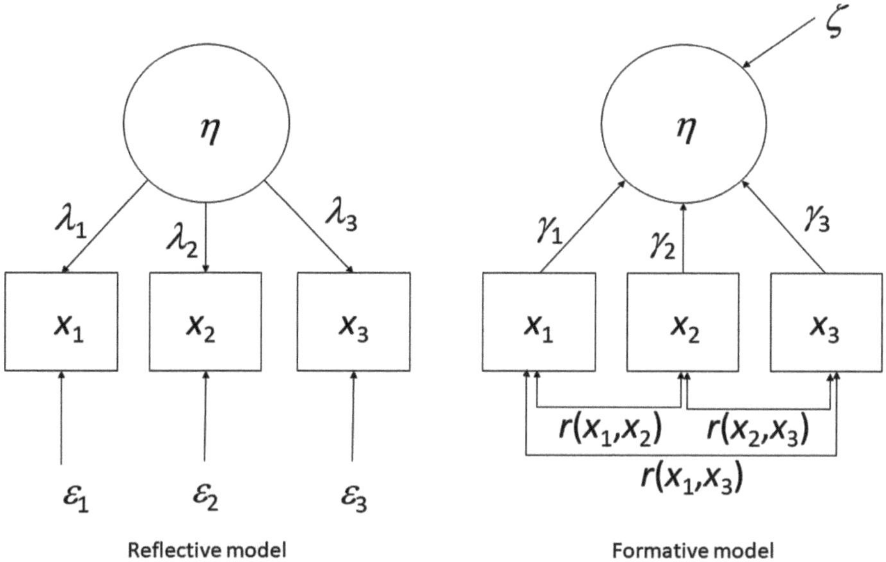

Fig. 2.1 Reflective model and formative model

to two different measurement approaches: *reflective* versus *formative* measurement models (Blalock 1964; Edwards and Bagozzi 2000; Diamantopoulos et al. 2008; Terzi et al. 2021). In the reflective model, the individual indicators are considered as a function of the latent variable, whereas in the formative model, the latent variable is considered as a function of the individual indicators.

Figure 2.1 shows the two different approaches graphically (Bollen and Lennox 1991; Mazziotta and Pareto 2017).

Traditionally, the reflective model is applied in the development of scaling models for subjective measurement, e.g. attitude or satisfaction scale construction (Jarvis et al. 2003), whereas the formative model is commonly used in the construction of composite indices based on both objective and subjective indicators (Maggino and Zumbo 2012). While the reflective view is predominant in psychological and management sciences, the formative view is common in economics and sociology (Coltman et al. 2008).

For the sake of simplicity, only linear models are considered here. However, nonlinear models can also be used.

2.2.1 Reflective Model

The most widely used approach is the *reflective* model (*top-down* approach), in which the individual indicators represent the effects (or manifestations) of an underlying latent variable (manifest variables). Therefore, causality is from the construct to the indicators, and a change in the phenomenon leads to variations in all its measures (i.e. the covariation among indicators reflects the change in the latent factor). In this model, the construct exists (in an absolute sense) independently of the researcher's perception or interpretation, even if it is not directly measurable (Borsboom et al. 2003). Specifically, the latent variable η represents the common cause shared by all indicators x_j that reflect the construct, with each indicator corresponding to a linear function of the underlying latent variable plus a measurement error:

$$x_j = \lambda_j \eta + \varepsilon_j \tag{2.1}$$

where x_j is indicator j, λ_j is a coefficient that captures the effect of η on x_j, and ε_j is the measurement error for indicator j. The measurement errors are assumed to be independent (i.e. $r(\varepsilon_i, \varepsilon_j) = 0$ for $i \neq j$) and unrelated to the latent variable (i.e. $r(\eta, \varepsilon_j) = 0$ for all j).[1]

A fundamental characteristic of reflective models is that the change in the latent variable must precede the change in the indicators. Thus, the indicators share a common theme and are interchangeable (adding or removing an indicator does not change the essential nature of the underlying construct). It follows that all indicators must be highly correlated with each other (i.e. all $r(x_i, x_j)$ with $i \neq j$ must have a high magnitude), and the correlations are explained by the measurement model.

Since high intercorrelations among indicators are desirable, measures such as factor loading and communality, Cronbach's alpha, average variance extracted, and internal consistency are used to empirically assess the individual and composite reliabilities of the indicators (Coltman et al. 2008).

Another important aspect that is often ignored concerns the polarity of the individual indicators. The 'polarity' of an individual indicator is the sign of the relationship between the indicator and the concept being measured. For example, in the case of well-being, the "Education rate" indicator has a positive polarity, while the "Unemployment rate" indicator has a negative polarity (see Sect. 3.2). In a reflective model, indicators with the same polarity must be positively correlated, while indicators with opposite polarity must be negatively correlated. Otherwise, the model leads to inconsistent results (Mazziotta and Pareto 2019; 2024).

Statistical approaches such as Factor Analysis and Principal Components Analysis[2] are often used to find sets of correlated indicators that are assumed to reflect underlying latent constructs (Shwartz et al. 2015).

---

[1] Note that formula (2.1) is a set of simple regression equations where the indicators are the dependent variables and the latent variable is the explanatory variable.

[2] It is important to stress that these two methods have different theoretical foundations. Factor Analysis is an explanatory model in which the observed variables (indicators) are assumed to be linear functions of a smaller number of unobserved variables (latent factors). Principal Component Analysis is a data reduction technique in which linear functions of the observed variables (indicators) are obtained in order to reproduce most of the variance with a smaller number of new variables (principal components). Therefore, the former hypothesizes an underlying latent structure of the variables, while the latter works without an explicit hypothesis (Krishnakumar and Nagar 2008).

A typical example of a reflective model is the measurement of a person's intelligence (Simonetto 2012). In this case, it is intelligence that influences responses and reaction times in an IQ test, and not vice versa. If a person's intelligence increases, the number of correct answers to all questions also increases and the reaction time decreases. Therefore, "Percentage of correct answers" has a positive polarity, while "Average response time" has a negative polarity. And these two indicators are negatively correlated.

Another example of a reflective model is the measurement of the corruption phenomenon, which is a latent not directly observable variable (Arbia 2018). In this case, a set of individual indicators selected on the basis of theoretical assumptions are designed as proxies for the phenomenon itself. The latent variable corruption η is the cause and the observed indicators x_j are the various effects of this cause, which remains hidden and unobservable.

A well-known reflective scale proposed in the literature is the Satisfaction With Life Scale (SWLS), which "*is narrowly focused to assess global life satisfaction*" (Diener et al. 1985). The scale is based on the idea that to measure the concept of life satisfaction, respondents must be asked for an overall assessment of their lives. It consists of five 7-point Likert items, and a person's score is the sum of the five items. A higher score means higher life satisfaction. The SWLS has been shown to have favourable psychometric properties, including high internal consistency and high temporal reliability.

2.2.2 Formative Model

The second approach is the *formative* model (*bottom-up* approach), in which the individual indicators are causes of an underlying latent variable, rather than its effects (causal variables). Therefore, causality is from the indicators to the construct, and a change in the phenomenon does not necessarily imply variations in all its measures (i.e. the latent variable is not assumed to explain the variances of the indicators and their covariation). In this model, the construct is defined by or is a function of the observed variables. It depends on an operationalist or instrumentalist interpretation by the researcher (Borsboom et al. 2003).

The specification of the formative model is as follows:

$$\eta = \sum_j \gamma_j x_j + \zeta \tag{2.2}$$

where γ_j is a coefficient capturing the effect of x_j on η, and ζ is an error term (disturbance).[3] ζ includes all remaining causes of the latent variable that are not represented in the indicators and are not correlated with them (i.e. $r(x_j, \zeta) = 0$),

[3] Note that formula (2.2) is a multiple regression equation where the latent variable is the dependent variable and the indicators are the explanatory variables.

while x_j is considered error-free.[4] Some authors omit the error term, so that (2.2) is reduced to a weighted linear combination of the x_j.

In this case, the change in the indicators must precede the change in the latent variable as its meaning depends on them, i.e. on the number and type of indicators. Thus, they do not have to share a common theme and are not interchangeable (adding or removing an indicator can change the underlying construct). It follows that indicators can have any intercorrelation pattern (i.e. all $r(x_i, x_j)$ with $i \neq j$ can be of different magnitude) and the correlations are not explained by the measurement model. Therefore, in a formative model, there is no relationship between polarities and correlations.

Since high intercorrelations among indicators are not desirable (to minimize multicollinearity), statistical approaches such as Factor Analysis and Principal Component Analysis or the examination of the correlation matrix for selecting indicators may overlook valid measures of the construct. Therefore, internal consistency is not relevant and no reliability test can be conducted for the indicators (Bollen and Lennox 1991).

Given that a formative model is not based on the assumption that the indicators are correlated, the correlation structure of the data cannot be used to determine the latent variable. Rather, the latent variable can be estimated by taking a weighted average (or other function) of the indicators that make up the construct (Shwartz et al. 2015).

A typical example of a formative model is the measurement of the well-being of a society. This phenomenon depends on health, income, employment, services, environment, etc., and not vice versa. If one of these factors improves, well-being increases (even if the other factors do not change). However, an increase in well-being does not necessarily go hand in hand with an improvement in all factors. For example, "GDP per capita" has a positive polarity, while "CO_2 emissions" has a negative polarity. However, these two indicators are not necessarily negatively correlated (in fact, they are usually positively correlated).

Another example of a formative model is the measurement of a person's socioeconomic status (SES), where indicators such as education, income, occupation and residence form the latent variable. If a person loses their job, this could have a negative impact on their SES. However, knowing that a person's SES has changed negatively does not mean that they have lost their job. Furthermore, a change in one indicator (income) does not necessarily imply a change in the same direction in the other indicators (education, occupation or residence) (Diamantopoulos and Winklhofer 2001).

One of the oldest and best-known formative composite indices in the literature is the UNDP's Human Development Index (HDI) (see Sect. 5.1). The index is a composite measure of human development that comprises three theoretical dimensions: health, knowledge and standard of living (UNDP 1990; 2010). Any change in one or more of these components is likely to lead to a change in a country's

[4] Although in reality this occurs very infrequently.

HDI score, but there is no reason to expect the components to correlate with each other. The same is true for the Canadian Index of Well-being (CIW), a composite measure of well-being based on eight pillars: community, democracy, education, environment, health, leisure and culture, standard of living and time use (Michalos et al. 2011) (see Sect. 5.3).

2.2.3 Differences Between Reflective and Formative Models

Latent variables are not all the same, they can have different characteristics and therefore require a different measurement model. Table 2.1 summarizes the main differences between the reflective and formative models[5]:

1. Nature of the construct. In a reflective model, the construct exists independently of the measures (the latent variable represents a real, thought independent entity), whereas in a formative model, the construct is defined by the researcher (the latent variable is conceived as an explanatory combination of its indicators).
2. Direction of causality. In a reflective model, causality flows from the construct to the indicators (a change in the construct leads to a change in the indicators), whereas in a formative model, causality flows from the indicators to the construct (a change in the indicators leads to a change in the construct under study).

Table 2.1 Reflective model versus formative model

Characteristic	Reflective model	Formative model
Nature of the construct	Existing, but not directly measurable	Defined by the researcher
Direction of causality	From construct to indicators	From indicators to construct
Indicators interchangeability	Yes	No
Indicators intercorrelation	Indicators should be correlated	Indicators can be incorrelated
Relationship between polarities and correlations	Indicators with equal polarities expected to be positively correlated, whereas indicators with opposite polarities expected to be negatively correlated	Polarities and correlations are independent
Error term	At the indicators level	At the construct level

[5] Several studies have also addressed mixed statistical modelling, incorporating both reflective and formative models (Jarvis et al. 2003; Diamantopoulos et al. 2008).

3. Indicators interchangeability. In a reflective model, the indicators all share a common theme and are interchangeable. This allows the researcher to measure the construct by sampling a few relevant indicators. In a formative model, the indicators define the construct and are not interchangeable. Thus, all indicators that form the latent variable should be included, i.e. a census of indicators is (ideally) required (redundant indicators should be eliminated a priori).
4. Indicators intercorrelation. In a reflective model, the indicators should be highly (positively or negatively) correlated, whereas in a formative model the indicators may be uncorrelated. Therefore, measures of *internal consistency* such as Cronbach's alpha can be used in a reflective model, but are inappropriate in a formative model where no theoretical assumption is made about the intercorrelation of the indicators.
5. Relationship between polarities and correlations. In a reflective model, indicators with equal polarities must be positively correlated, while indicators with opposite polarities must be negatively correlated. In a formative model, polarities and correlations are independent. This point is particularly important because methods such as Factor Analysis and Principal Component Analysis aggregate indicators based on their correlations, regardless of their polarities.
6. Error term. In a reflective model, all error terms are associated with the observed indicators and therefore represent measurement error in the latent variable. In a formative model, the error term is not associated with the individual indicators and therefore does not represent measurement error. It can be eliminated by capturing all possible causes for the construct or by specifying the focal construct in such a way that the entire set of indicators is captured (Diamantopoulos 2006).

It is important to stress that the procedures for developing measures associated with the two approaches are very different. The former place major emphasis on the intercorrelations among the individual indicators, focusing on common variance, and emphasize unidimensionality and *internal consistency*. The latter focus on explaining abstract (unobserved) variance and emphasize the role of indicators as predictors rather than predicted variables (Diamantopoulos and Siguaw 2006).

Since the causal relationship and correlation play a fundamental role in the specification of the model, it is useful to shed more light on the difference between these two terms.

Figure 2.2 illustrates the difference between causality and correlation using a cartoon bubble. Causality implies a direct cause-and-effect relationship, whereas correlation does not establish causality but a simple covariation. Causality implies a temporal sequence in which the cause precedes the effect. Correlation, on the other hand, does not require a specific temporal sequence between the variables and measures their relationship.

In the figure, a pick-up lorry has broken down and the driver asks for help to push it to the nearest garage. Two men are pushing: one on the pick-up, who pushes the cab with his hands without obviously having any effect on the vehicle's progress; a second, who is on the road, pushes the entire pick-up, causing the

Fig. 2.2 Causality and correlation

vehicle to move forwards. In the first case, the person with the red shirt pushes in the same direction but cannot move the vehicle forwards, representing a correlation (there is no effect because there is no cause); in the second case, the person with the light blue shirt pushes in the same direction and also enables the vehicle to move, representing a causality. There is therefore a 'correlation' between the movement of the person with the red shirt and the movement of the pick-up (because they both move in the same direction), while there is a 'causality' between the movement of the person with the light blue shirt and the movement of the pick-up (because the first pushes the second in the same direction).

Correlation represents the relationship between two variables/indicators. If the variables show 'parallel' trends, they are correlated, but this does not necessarily mean that one is the cause of the other. This is why it is said that: "Correlation is not necessarily an indication of causality". A strong correlation could be an indication of causality, but there could also be other explanations. For example, variables may be correlated but in reality there is no underlying relationship (spurious correlation) or there may be a hidden third variable (latent factor) that makes the relationship appear stronger (or weaker) than it is.

2.2.4 Misspecification of the Model

The choice of measurement model should be theoretically driven, i.e. based on the 'auxiliary theory' that specifies the nature and direction of the relationship between construct and measures (Edwards and Bagozzi 2000). In many cases, this choice is straightforward because the causal priority between the construct and the indicators is very clear. For example, intelligence is typically modelled as reflective, whereas SES is usually modelled as formative. In other cases, however, the choice between a reflective and a formative measurement model may not be so easy, as "*the directionality of the relationship is far from obvious*" (Fayers and Hand 1997). Furthermore, it is not always possible to determine a priori whether a particular construct is inherently reflective or formative. It is often the structure of the observed measures that determines the nature of the construct, and some constructs can be modelled as both reflective and formative (Wilcox et al. 2008; MacKenzie et al. 2011).

Several methodological studies have provided examples of constructs that were erroneously operationalized using reflective models when a formative perspective would have been theoretically appropriate.

The choice between reflective and formative models has a significant impact on outcomes, but is hardly recognized in the literature (Jarvis et al. 2003). Most researchers do not address the issue and apply the procedures mechanically without questioning whether or not they are appropriate for the concept being measured. On the other hand, one of the best-known texts on this topic, the OECD's "Handbook on Constructing Composite Indicators. Methodology and User Guide" (2008), makes no reference to the measurement model. In this regard, some authors speak of an "*almost automatic acceptance of the reflective model in the minds of researchers*" (Diamantopoulos and Winklhofer 2001).

Consequently, model misspecification commonly consists of adopting a reflective model where a formative approach would be appropriate (Diamantopoulos and Siguaw 2006). The other case of misspecification, i.e. the incorrect adoption of a formative model when a reflective approach would be appropriate, is rather negligible. One possible explanation for this is that procedures for developing and evaluating measures for reflective models have a long tradition in the social sciences and have been established over the years, whereas concrete guidelines for the construction of formative indices have only recently been proposed (Diamantopoulos and Winklhofer 2001). Furthermore, some researchers are simply unaware of the conceptual distinction between formative and reflective measurement models, although this distinction was made in the literature almost 50 years ago (Jarvis et al. 2003).

Avoiding misspecification is crucial to prevent false conclusions from empirical analyses. Incorrect specification of the measurement model when constructing a composite index can seriously affect the results. Therefore, the researcher must think carefully about the direction of causality between the construct being measured and the individual indicators.

A typical example of misspecification is the construction of a composite index based on Principal Component Analysis (PCA) when a formative model is required (Mazziotta and Pareto 2019; 2024).

As is well known, the definition of the principal component as the weighted sum of the individual indicators (see Formula 3.9) is similar to Eq. (2.2) and therefore suggests that a composite index based on PCA can be used with a formative measurement model (Edwards and Bagozzi 2000; Zumbo 2007; Markus and Borsboom 2013).

However, Nunnally and Bernstein (1994) show that the equations of PCA can be written with both the principal components as a function of the observed variables (similar to a formative model) and the observed variables as a function of the principal components (similar to a reflective model) without loss of meaning. They refer to this as a *component* model and emphasize that most conceptions of factors in psychological research imply a reflective or component model, as the component model can be seen as a special case of Factor Analysis (FA) in which the variables are reproduced without error.

On the other hand, Bollen and Lennox (1991) note that Eq. (2.2) differs from a principal component model in that there is a disturbance term and only one latent variable for a set of indicators, whereas PCA yields a number of principal components equal to the number of individual indicators.

In addition, some important aspects must be considered:

1. In a PCA-based composite index (e.g. the first principal component), the weights depend on the correlations between indicators, as they are given by the eigenvectors of the correlation (or covariance) matrix. However, correlations between individual indicators are not relevant in a formative model and cannot be explained by it, as the indicators do not necessarily share the same theme and therefore do not have a 'preconceived' pattern of intercorrelation (Coltman et al. 2008). In this regard, Fayers and Hand (2002) pointed out in an article in the "Journal of the Royal Statistical Society" that the main focus of methods of scale construction and assessment of standard psychometric approaches "*has been the correlation structure of the data – which is inappropriate for causal variables*". And they added that, when causal variables are involved, "*it is disturbing to note that anyone developing a scale by using traditional methods would remain blissfully unaware that they may be omitting important items or including inappropriate items*". The idea of PCA is to account for the greatest possible variation of the individual indicators with the smallest possible number of components (see Sect. 3.3.3). Therefore, in PCA, the weights are only used to correct for overlaps between two or more correlated indicators and are not a measure of the theoretical importance of the associated indicators (OECD 2008). If no correlation is found between the individual indicators, the weights cannot be estimated using this method.
2. By construction, a PCA index assigns larger weights to highly correlated indicators (because they help to maximize the sum of the squared correlation coefficients between the index and the individual indicators). However, in a multiple regression, such as that in formula (2.2), the individual indicators should have little or no correlation with each other in order to avoid multicollinearity. Thus, while in reflective models each individual indicator is by definition collinear with the others, multicollinearity in formative models is harmful and can lead to unstable weights (Tabachnick and Fidell 2001). In addition, excessive multicollinearity makes it difficult to separate the different influence of the individual indicators on the latent variable (Bollen 1989; Diamantopoulos and Winklhofer 2001). Therefore, collinearity among individual indicators challenges the interpretation of formative composite indices (Cenfetelli and Bassellier 2009).

3. Under certain conditions, the principal components are equivalent to the factor scores obtained by Factor Analysis (FA) and can then be considered as estimators of latent factors (Krishnakumar and Nagar 2008). For example, in cases with a large number of variables (e.g. 30 or more) and/or high estimated communality, FA and PCA yield similar results (Gorsuch 1983; Hair et al. 2006). Moreover, both types of analysis have a common mathematical core, are based on the correlation (or covariance) matrix of the data, and use matrix decomposition methods to obtain the components or factors. When 'principal components' is used as the method for factor extraction, the matrix of factor loadings obtained in FA is identical to the matrix of correlation coefficients between original variables and principal components obtained in PCA. Hence, the factors before rotation are identical to the first few principal components with the largest variances (Gniazdowski 2017, 2021). However, FA is based on a reflective measurement model, so PCA cannot be considered truly formative.

From the above, it is clear that PCA is much closer to the reflective model than the formative model.

Indeed, PCA is considered an appropriate method for examining the latent structure underlying a set of indicators (Bohrnstedt 1970; Vinzi et al. 2003), and the use of PCA to test the *content validity* or *construct validity* of reflective models can be found in several studies in the literature since the 1970s, particularly in psychological and clinical fields.

References

Arbia G (2018) Aspetti statistico-metodologici nella costruzione di indicatori di corruzione e nella valutazione della loro efficacia. In: Gnaldi M, Ponti B (eds) Misurare la corruzione oggi. Franco Angeli, Milan

Blalock HM (1982) Conceptualization and measurement in the social sciences. Sage Publications, Beverly Hills, Calif

Blalock HM (1964) Causal inferences in nonexperimental research. University of North Carolina Press, Chapel Hill, NC

Blalock HM (1968) The measurement problem: a gap between the languages of theory and research. In: Blalock HM, Blalock AB (eds) Methodology in social research. McGraw-Hill, New York

Bohrnstedt GW (1970) Reliability and validity assessment in attitude measurement. In: Summers GF (ed) Attitude measurement. Rand McNally, London, pp 80–99

Bollen K (1989) Structural equations with latent variables, 1st edn. Wiley-Interscience, Malden

Bollen KA, Lennox R (1991) Conventional wisdom in measurement: a structural equation perspective. Psychol Bull 110:305–314

Borsboom D, Mellenbergh GJ, Heerden JV (2003) The theoretical status of latent variables. Psychol Rev 110:203–219

Cenfetelli RT, Bassellier G (2009) Formative measurement in information systems research. MIS Q 33:689–707

Coltman T, Devinney TM, Midgley DF, Venaik S (2008) Formative versus reflective measurement models: two applications of formative measurement. J Bus Res 61:1250–1262

Diamantopoulos A (2006) The error term in formative measurement models: interpretation and modeling implications. J Model Manag 1:7–17

Diamantopoulos A, Siguaw JAA (2006) Formative versus Reflective indicators in organizational measure development: a comparison and empirical illustration. Br J Manag 17:263–282
Diamantopoulos A, Winklhofer HM (2001) Index construction with formative indicators: an alternative to scale development. J Mark Res 38:269–277
Diamantopoulos A, Riefler P, Roth KP (2008) Advancing formative measurement models. J Bus Res 61:1203–1218
Diener E, Emmons RA, Larsen RJ, Griffin S (1985) The satisfaction with life scale. J Pers Assess 49:71–75
Edwards JR, Bagozzi RP (2000) On the nature and direction of relationships between constructs and measures. Psychol Methods 5:155–174
Fayers PM, Hand DJ (1997) Factor analysis, causal indicators and quality of life. Qual Life Res 6:139–150
Fayers PM, Hand DJ (2002) Causal variables, indicator variables and measurement scales: an example from quality of life. J R Stat Soc A 165:233–261
Gniazdowski Z (2017) New interpretation of principal components analysis. Zeszyty Naukowe WWSI 11:43–65
Gniazdowski Z (2021) Principal component analysis versus factor analysis. Zeszyty Naukowe WWSI 15:35–88
Gorsuch RL (1983) Factor analysis, 2nd edn. Lawrence Erlbaum Associates, Hillsdale
Hair Jr JF, Black WC, Babin BJ, Anderson RE, Tatham RL (2006) Multivariate data analysis (6th ed.). Pearson Prentice Hall, Upper Saddle River
Jarvis CB, Mackenzie SB, Podsakoff PM (2003) A critical review of construct indicators and measurement model misspecification in marketing and consumer research. J Consum Res 30:199–218
Krishnakumar J, Nagar AL (2008) On exact statistical properties of multidimensional indices based on principal components, factor analysis, MIMIC and structural equation models. Soc Indic Res 86:481–496
MacKenzie SB, Podsakoff PM, Podsakoff NP (2011) Construct measurement and validation procedures. MIS Q 35:293–334
Markus KA, Borsboom D (2013) Frontiers of test validity theory. Measurement, causation, and meaning. Routledge, New York
Mazziotta M, Pareto A (2017) Synthesis of indicators: the composite indicators approach. In: Maggino F (ed) Complexity in society: from indicators construction to their synthesis. Springer, Cham, pp 159–191
Maggino F, Zumbo BD (2012) Measuring the quality of life and the construction of social indicators. In: Handbook of social indicators and quality of life research. Springer, Dordrecht, pp 201–238
Michalos AC, Smale B, Labonté R, Muharjarine N, Scott K, Moore K et al (2011) The Canadian index of wellbeing. Technical report 1.0. Canadian Index of Wellbeing and University of Waterloo, Waterloo, ON
Mazziotta M, Pareto A (2019) Use and misuse of PCA for measuring well-being. Soc Indic Res 142:451–476
Mazziotta M, Pareto A (2024) Principal component analysis for constructing socio-economic composite indicators: theoretical and empirical considerations. SN Soc Sci 4:114. https://doi.org/10.1007/s43545-024-00920-x
Nunnally JC, Bernstein IH (1994) Psychometric Theory, 3rd edn. McGraw-Hill, New York
OECD (2008) Handbook on constructing composite indicators. Methodology and user guide. OECD Publications, Paris
Shwartz M, Restuccia JD, Rosen AK (2015) Composite measures of health care provider performance: a description of approaches. Milbank Q 93:788–825
Simonetto A (2012) Formative and reflective models: State of the art. Electronic Journal of Applied Statistical Analysis 5:452–457
Tabachnick BG, Fidell LS (2001) Using multivariate statistics, 4th edn. HarperCollins, New York

Terzi S, Otoiu A, Grimaccia E, Mazziotta M, Pareto A (2021) Open issues in composite indicators. A starting point and a reference on some state-of-the-art issues. Roma TrE-Press, Rome
UNDP (1990) Human development report 1990: concept and measurement of human development. Oxford University Press, Oxford
UNDP (2010) Human development report 2010. Palgrave Macmillan, New York
Vinzi VE, Lauro C, Tenenhaus M (2003) PLS path modeling. Working Paper. DMS—University of Naples, HEC—School of Management, Jouy-en-Josas
Wilcox JB, Howell RD, Breivik E (2008) Questions about formative measurement. J Bus Res 61:1229–1237
Zumbo BD (2007) Validity: foundational issues and statistical methodology. In: Rao CR, Sinharay S (eds) Handbook of statistics. Elsevier, Boston, pp 45–79
Zeller RA, Carmines EG (1980) Measurement in the social sciences. The link between theory and data. Cambridge University Press, New York

Composite Indicators Construction

3

We have seen that the most important steps in the construction of a composite indicator are as follows[1]: (1) Definition of the phenomenon, (2) Selection of the individual indicators, (3) Exploratory analysis, (4) Normalization of the individual indicators, (5) Aggregation of the normalized indicators, (6) Validation of the composite index, and (7) Presentation of the results. This chapter describes the characteristics of each step and the main tools available.

3.1 Definition of the Phenomenon

The definition of the phenomenon should provide a clear and concise idea of what the composite index is intended to measure. The phenomenon should be defined positively (not by negating other things), in unambiguous terms and in a way that is consistent with previous research. The definition should not be too technical, allow for multiple interpretations, be circular, tautological or self-referential. In particular, the researcher should specify the nature of the phenomenon, including (i) the type of property the phenomenon represents (e.g. a perception, an outcome, a performance), and (ii) the entity (*unit*) to which it refers (e.g. a person, an organization, a country) (MacKenzie et al. 2011).

After carefully defining the phenomenon, it is important to examine all of its distinguishable facets or constituent aspects (*dimensions* or *pillars*) and define each of them with the same care that was applied in the case of the phenomenon itself.

[1] Some authors describe a larger number of steps (e.g. the imputation of missing data). In this chapter, only the basic steps required to create a composite indicator are described.

M. Mazziotta and A. Pareto, *Statistics for Composite Indicators*, SxI - Springer for Innovation / SxI - Springer per l'Innovazione 18,
https://doi.org/10.1007/978-3-031-97978-1_3

These dimensions can in turn be subdivided into sub-dimensions (*components*). Finally, each dimension or sub-dimension must be represented by an individual indicator that is measured empirically. This scheme is illustrated graphically in Fig. 3.1.

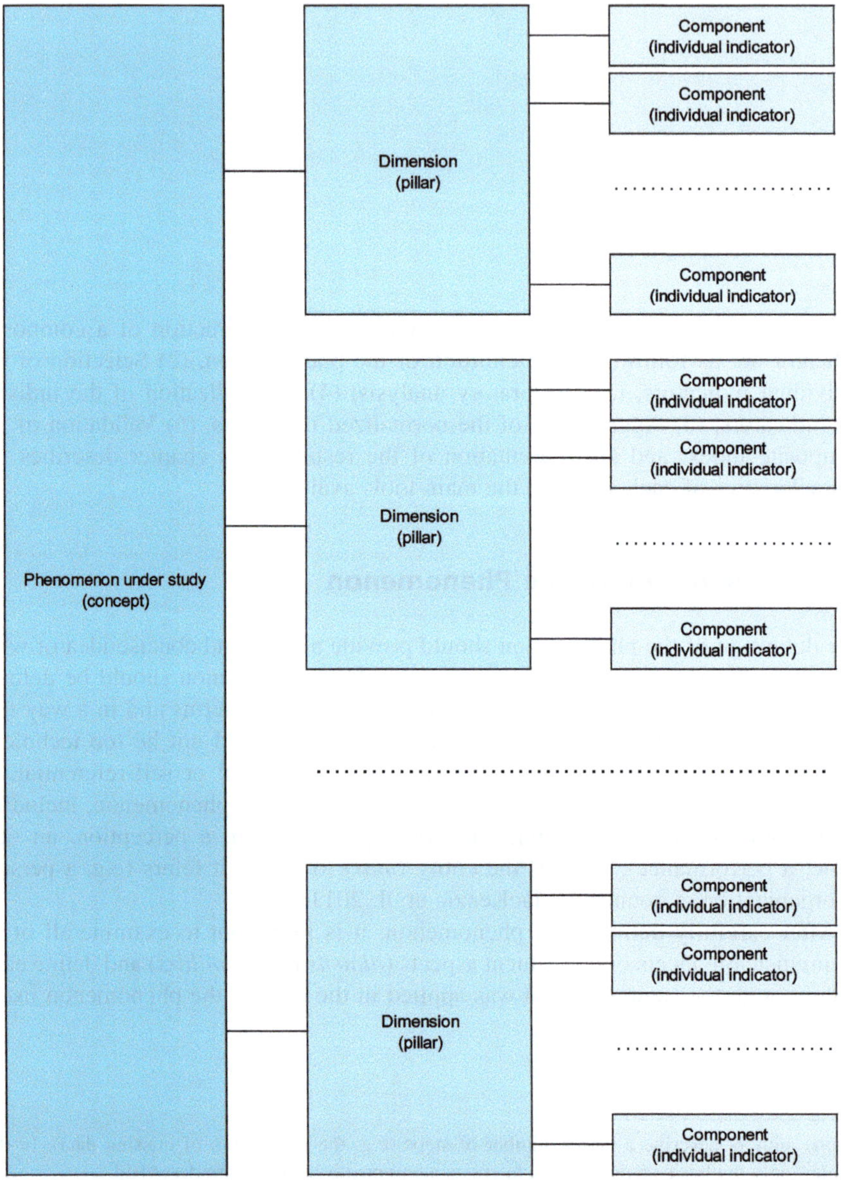

Fig. 3.1 Conceptual scheme of the phenomenon under study

A fundamental aspect concerns the identification of the measurement model, for which—as we know—two different approaches can be distinguished (Diamantopoulos et al. 2008):

- reflective model, when the indicators are considered as an 'effect' of the phenomenon to be measured, so that a change in the latent variable is reflected in a change in the observed indicators (the indicators are interchangeable and the correlations between them are explained by the model);
- formative model, when the indicators are considered as the 'cause' of the phenomenon to be measured, so that a change in the latent variable does not necessarily imply a change in all the observed indicators (the indicators are not interchangeable and the correlations between them are not explained by the model).

The definition of the measurement model is very important as it is closely linked to the selection and aggregation of the individual indicators.

3.2 Selection of the Individual Indicators

In this step, it is necessary to determine the number and nature of the components of the phenomenon to be measured. If the measurement model is reflective, we must select manifest 'variables'; if the model is formative, we must select causal 'variables'.

The selected indicators must be static or dynamic. Static indicators can be of different types:

- *Composition ratios*, obtained by dividing the frequency/amount of a category of a variable (a_i) by the total frequency/amount of the variable, i.e. by the sum of all possible categories ($a = a_1 + a_2 + \cdots + a_k$) (often multiplied by 100):

$$\frac{a_i}{a_1 + a_2 + \cdots + a_n} = \frac{a_i}{a}$$

  Examples are percentage indicators such as "Percentage of adults literate", "Percentage of children enrolled in school" and "Percentage of labour force in agriculture".
- *Derivation ratios*, obtained by dividing the frequency/amount of a category of a variable (a_i) by the frequency/amount of a category of another variable (b_i) that is its cause or its necessary antecedent or its logical precondition (often multiplied by 100 or more if necessary):

$$\frac{a_i}{b_i}$$

  Example are "Infant mortality rate", "Fertility rate" and "Incidence rate of occupational injuries".

- *Density ratios*, obtained by dividing the total frequency/amount of a variable (a) by the total frequency/amount of another variable to which it refers (b) (sometimes multiplied by 100 or more):

$$\frac{a}{b}$$

  Examples are per capita indicators such as "GDP per capita" and "Energy consumption per capita" as well as inverse per capita indicators such as "Population density" and "Inhabitants per doctor".
- *Coexistence ratios*, obtained by dividing the frequency/amount of a category of a variable (a_i) by the frequency/amount of another category of the same variable (a_j) (often multiplied by 100):

$$\frac{a_i}{a_j}$$

  Example are "Gender ratio", "Dependency ratio" and "Imports to exports ratio".

Dynamic indicators are usually expressed as a relative difference of the following type (often multiplied by 100):

$$\frac{a_1 - a_0}{a_0}$$

where a_0 and a_1 are the frequencies/amounts of a variable at time 0 and 1 respectively. A typical example is "GDP growth rate".

For each component of the phenomenon, an individual quantitative indicator (discrete or continuous) that measures it empirically must be identified and its polarity defined.

The polarity of an individual indicator is the sign of the relationship between the indicator and the phenomenon to be measured (Mazziotta and Pareto 2017). For example, if the phenomenon to be measured is development, the indicator "Life expectancy" has a positive polarity, as development increases with increasing life expectancy, while the indicator "Infant mortality rate" has a negative polarity, as development decreases with increasing infant mortality. If, on the other hand, the phenomenon to be measured is poverty, the "Life expectancy" indicator has a negative polarity, as poverty decreases with increasing life expectancy, while the "Infant mortality rate" indicator has a positive polarity, as poverty increases with increasing infant mortality. The polarity of an individual indicator is therefore a variable concept that depends on the nature of the phenomenon under investigation.

In general, the choice of individual indicators is based on theoretical and empirical aspects, pragmatism and intuition (Booysen 2002).

The strengths and weaknesses of a composite index reflect the quality of the underlying individual indicators. The individual indicators should be selected on the basis of their relevance, validity, availability, timeliness, etc. (OECD 2008). For example, if the composite index is to be calculated annually, individual indicators that are available over several years cannot be included. If the composite index is to be calculated by region, individual indicators that are available at national level cannot be taken into account. It should also be noted that equally relevant individual indicators must be weighted equally in the aggregation step, while indicators with different relevance require the definition of an 'ad hoc' weighting system.

The selection step is the result of a trade-off between possible redundancies and loss of information. Redundancies can be evaluated using 'correlations' while the loss of information is measured using the 'variance explained'. Therefore, a statistical approach for selecting indicators is to perform a principal component analysis (see Sect. 3.3.3) of the potential indicators and include or exclude indicators based on their correlations with the components. However, the selection process depends on the measurement model chosen: in a reflective model, all indicators must be correlated with each other, whereas in a formative model they can be correlated or uncorrelated (Diamatopoulos et al. 2008). In particular:

- in a reflective model, if the variance explained by the first principal component is satisfactory (e.g. 80%), a single composite index can be created containing the indicators that correlate most strongly with the first principal component. If the variance explained by the first principal component is not satisfactory (e.g. 50%), it is more appropriate to construct a composite index for each significant principal component that includes the indicators most highly correlated with the corresponding principal component;
- in a formative model, the variance explained and the number of significant principal components are less important. If we want to construct a composite index that includes the indicators that are less correlated with each other (Salzman 2003), we can select the indicator that is most correlated with each significant principal component. However, it is important to remember that individual indicators can also be correlated with each other.

Although the presence of highly correlated individual indicators is not a prerequisite for the formative model (as they generate multicollinearity), interestingly some authors test the internal consistency of the formative indicators and exclude from the analysis those that are not consistent with the others (Annoni and Bolsi 2020). According to Bollen and Lennox (1991), this practice "*can have dire consequences*" (p. 307) (see Sect. 2.2.2).

3.3 Exploratory Analysis

Exploratory analysis is a series of activities carried out to investigate the distribution and characteristics of each indicator and the relationships between them. This step, which is often underrated, is fundamental as it allows the researcher to fully understand the characteristics of the indicators to be normalized and aggregated and can therefore provide useful indications on the methods to be used in the subsequent steps. The analysis also makes it possible to determine whether it is necessary to perform preliminary transformations of the individual indicators.

3.3.1 Univariate Analyses

In univariate analyses, the individual indicators are analysed individually, i.e. one at a time, in order to examine the shape of the frequency distribution and calculate the main descriptive statistics.

Given a set of m individual indicators for n statistical units, let x_{ij} be the value of indicator j for unit i. Then, for indicator j, we have:

- Minimum and maximum

$$\mathrm{Min}_{x_j} = \min_i (x_{ij})$$
$$\mathrm{Max}_{x_j} = \max_i (x_{ij})$$

  where $\min_i (x_{ij})$ and $\max_i (x_{ij})$ are the minimum and maximum of indicator j respectively. Min and Max allow us to identify incorrect or anomalous values (*outliers*), i.e. values that deviate significantly from other observations. The difference between Max and Min is the distribution range.
- Mean and median

$$\mathrm{M}_{x_j} = \frac{1}{n}\sum_{i=1}^{n} x_{ij}$$
$$\mathrm{Med}_{x_j} = \underset{i}{\mathrm{median}} (x_{ij})$$

  where $\underset{i}{\mathrm{median}} (x_{ij})$ is the central value (if n is odd) or the half-sum of the two central values (if n is even) in the set of ordered values of indicator j. M and Med are position indices of the distribution and allow an approximate determination of the degree of symmetry. If M = Med, the distribution is symmetrical (no skewness); if M > Med, the distribution is right-skewed (positive skewness); if M < Med, the distribution is left-skewed (negative skewness).
- Standard deviation and coefficient of variation

$$\mathrm{S}_{x_j} = \sqrt{\frac{1}{n}\sum_{i=1}^{n} (x_{ij} - \mathrm{M}_{x_j})^2}$$
$$\mathrm{CV}_{x_j} = \mathrm{S}_{x_j}/\mathrm{M}_{x_j}$$

  S and CV are variability indices of the distribution and make it possible to assess the degree of dispersion. The higher the value of the indices, the greater the dispersion of the distribution. S is expressed in the same unit of measurement as the individual indicator, and the square of S gives the *variance*, which is indicated by S^2. CV is a pure number that can be used to compare the variability of indicators expressed in different units of measurement and to assess the impact of so-called 'implicit weighting' (see Sect. 3.5.1).

- Skewness and kurtosis indices

$$\gamma_{1x_j} = \frac{1}{n}\sum_{i=1}^{n}\left(\frac{x_{ij} - \mathrm{M}_{x_j}}{\mathrm{S}_{x_j}}\right)^3$$

$$\gamma_{2x_j} = \frac{1}{n}\sum_{i=1}^{n}\left(\frac{x_{ij} - \mathrm{M}_{x_j}}{\mathrm{S}_{x_j}}\right)^4 - 3$$

γ_1 and γ_2 are shape indices of the distribution. If $\gamma_1 = 0$, there is symmetry; if $\gamma_1 > 0$, there is positive skewness; if $\gamma_1 < 0$, there is negative skewness. The kurtosis index is relevant only in the case of unimodal frequency distributions, whose curve is bell-shaped. In this case, if $\gamma_2 = 0$, the curve is mesokurtic (similar to a normal one); if $\gamma_2 > 0$, the curve is leptokurtic (more pointed than a normal curve); if $\gamma_2 < 0$, the curve is platykurtic (flatter than a normal curve).

In any case, it is advisable to use graphical representations such as the frequency histogram and the boxplot for a precise examination of the shape of the distribution. The boxplot in particular allows us to compare different distributions simultaneously. With a large number of individual indicators expressed in the same unit of measurement, it is therefore possible to use a single boxplot instead of resorting to many frequency histograms. On the other hand, the boxplot does not provide information about the mode and therefore does not allow the identification of multimodal distributions.

As is well known, most traditional statistical methods are based on the assumption of normality of the variables under consideration. However, since most indicators used to construct composite indices are fractions, they are rarely normally distributed.

For example, there are two main types of socioeconomic indicators:

1. Per capita indicators that have no fixed upper limit (e.g. "GDP per capita") and tend to have positively skewed distributions with some outliers.
2. Percentage indicators that have a fixed range of variation (e.g. "Education rate") and tend to have distributions with negative skewness and a high concentration of observations close to the upper limit.

These characteristics can lead to a distortion of the results and must be carefully checked when normalizing the indicators (see Sect. 3.4.3). The presence of skewed distributions also affects the bivariate relationships between individual indicators, which are often nonlinear.

In this case, it may be useful to transform the individual indicators accordingly. For example, if an indicator has a logarithmic normal distribution, it can be brought into the normal distribution by a logarithmic transformation.

A more general transformation that includes both logarithms and exponentiations is the Box-Cox transformation shown below (Box and Cox 1964). However, it should be noted that these transformations change the 'distances' between the statistical units and thus the correlations between the transformed indicators.

Box-Cox Transformation

This is a transformation that makes it possible to make the distributions more symmetrical and to linearize the relationships between the variables. The transformation assumes that all values of the indicator are greater than zero and has the form:

$$x'_{ij} = \frac{x_{ij}^{\lambda}-1}{\lambda} \quad \text{if } \lambda \neq 0$$
$$x'_{ij} = \log(x_{ij}) \text{ if } \lambda = 0$$

where λ is a parameter that depends on the shape of the distribution.

The Box-Cox transformation includes several special cases. For example, for $\lambda = -1$ we have the reciprocal, for $\lambda = -0.5$ the reciprocal of the square root, for $\lambda = 0.5$ the square root and for $\lambda = 2$ the squaring. The transformation is continuous and monotonically increasing. It is concave if $\lambda < 1$ and convex if $\lambda > 1$. Some examples can be found in Fig. 3.2.

Due to these properties, the Box-Cox transformation causes a contraction of the highest values when $\lambda < 1$ and an elongation of the highest values when $\lambda > 1$. Thus, if the distribution is right-skewed, we must choose a negative value for λ, if it is left-skewed, we must choose a positive value for λ.

If the distribution of the original indicator is unimodal, the value of λ that best approximates the normal distribution for the transformed indicator is usually selected.

3.3.2 Bivariate Analyses

Bivariate analyses aim to investigate the relationships between pairs of individual indicators. Measures of the strength of association between two variables are used to examine the relationship, while regression methods are used to examine the covariation. These analyses should be descriptive, unless the statistical units analysed are a probabilistic sample. For socioeconomic indicators, there are usually only a limited number of countries, regions or provinces, so the units do not represent a sample of a larger population.

As far as the graphical representation is concerned, the scatterplot is used.

Association analysis

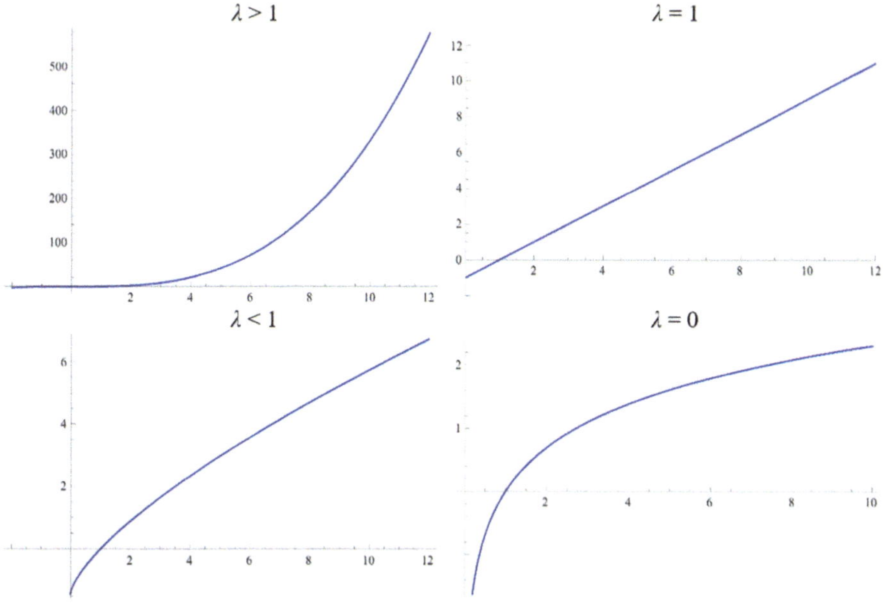

Fig. 3.2 Box-Cox transformation

Below are some measures for analysing the association between individual indicators.

1. *Bravais-Pearson linear correlation coefficient*

This is the most commonly used association index and measures the degree of linear relationship between two variables. Given two indicators j and k, we have:

$$r_{x_j,x_k} = \frac{\sum_{i=1}^{n}(x_{ij} - \mathrm{M}_{x_j})(x_{ik} - \mathrm{M}_{x_k})}{\sqrt{\sum_{i=1}^{n}(x_{ij} - \mathrm{M}_{x_j})^2 \sum_{i=1}^{n}(x_{ik} - \mathrm{M}_{x_k})^2}}$$

r is a pure number and varies between -1 and 1. If $r = 0$, the indicators are uncorrelated (there is no linear relationship); if $r > 0$, the indicators are positively correlated (if one increases, the other also increases); if $r < 0$, the indicators are negatively correlated (if one increases, the other decreases). Furthermore, if $r = 1$, there is a maximum positive correlation (all observations of the bivariate distribution lie on a line with a positive slope); if $r = -1$, there is a maximum negative correlation (all observations lie on a line with a negative slope).

The index can be seen as a measure of the goodness of fit of conventional regression lines (the regression line of indicator j as a function of indicator k and vice versa). However, it is not very robust and can lead to extremely biased results if there is even a single outlier in the bivariate distribution. In this case, it may indicate a low degree of association even in the presence of a strong linear relationship or vice versa. If the indicators used are not normally distributed, it is also less efficient than other measures.

2. *Spearman rank correlation coefficient*

This is a rank-based index that measures the degree of agreement between rankings according to two variables. The formula is as follows:

$$\rho_{x_j,x_k} = 1 - \frac{6\sum_{i=1}^{n}(\text{rank}(x_{ij}) - \text{rank}(x_{ik}))^2}{n(n^2-1)}$$

where $\text{rank}(x_{ij})$ is the position (rank) of unit i according to indicator j.

ρ varies between -1 and 1. If $\rho = 0$, there is no correlation between the ranks; if $\rho > 0$, there is a positive correlation between the ranks; if $\rho < 0$, there is a negative correlation between the ranks. Furthermore, if $\rho = 1$, there is a perfect positive correlation between the ranks (each unit occupies the same position in both rankings); if $\rho = -1$, there is a perfect negative correlation between the ranks (a unit is in first place in one ranking and in last place in the other, etc.).

The index is a special case of the Bravais-Pearson correlation coefficient, where the original values are replaced by the corresponding ranks (see point 2 in Sect. 3.4.2).

As it is based on ranks, ρ is a more robust index than r and also enables the measurement of nonlinear relationships between indicators.[2]

3. *Correlation Coefficient based on the L_1-norm*

This is an index that was proposed as an alternative to the linear correlation coefficient of Bravais-Pearson for the study of relationships between socioeconomic indicators (McGranahan et al. 1985). It is also referred to as *absolute ratio correlation* (Pareto 2023) and has the following form:

$$\begin{cases} r'_{x_j,x_k} = 1 - \dfrac{\sum_{i=1}^{n}\left|x_{ij} - \hat{x}_{ij}^{+}\right|}{\sum_{i=1}^{n}\left|x_{ij} - \hat{x}_{ij}^{-}\right|} & \text{if } \left|x_{ij} - \hat{x}_{ij}^{+}\right| \leq \left|x_{ij} - \hat{x}_{ij}^{-}\right| \\ r'_{x_j,x_k} = \dfrac{\sum_{i=1}^{n}\left|x_{ij} - \hat{x}_{ij}^{-}\right|}{\sum_{i=1}^{n}\left|x_{ij} - \hat{x}_{ij}^{+}\right|} - 1 & \text{if } \left|x_{ij} - \hat{x}_{ij}^{+}\right| > \left|x_{ij} - \hat{x}_{ij}^{-}\right| \end{cases}$$

where $\hat{x}_{ij}^{+} = a^{+} + bx_{ik}$ and $\hat{x}_{ij}^{-} = a^{-} - bx_{ik}$, with $b > 0$, are the estimates of indicator j as a function of indicator k, using two median lines with opposite slopes that interpolate the bivariate distribution so as to maximize the absolute value of

---

[2] Note that the ranking of the data converts a nonlinear strictly monotonic relationship into a linear relationship.

r'.[3] r' also varies between -1 and 1 and can be interpreted in the same way as the classical correlation coefficient.

The index can be seen as a measure of the goodness of fit of a median line, called *best fitting line* (BFL), which describes the linear relationship between the two indicators, with respect to the median line with the opposite slope, called *counter line* (CL), which describes the opposite relationship.

As it is based on absolute deviations, r' is a very robust index in terms of outliers and is particularly suitable for indicators with skewed distributions, such as socioeconomic indicators.

Covariation analysis

The study of covariation between indicators is particularly important and is often performed independently of the construction of a composite index.

In general, a simple linear regression based on the least squares method is used. This method allows the construction of the regression line of one indicator as a function of the other and is based on the assumption that the data for the independent variable x are exact, while the data for the dependent variable y show deviations of a random nature.

However, in the case of indicators, especially socioeconomic ones, the traditional concept of regression of one variable on another and the hypothesis of random errors in only one of the two variables considered have no rational justification. Identifying one variable as independent (without error) and another as dependent (with error) is typical of experimental studies, but cannot be justified in observational studies.[4]

For example, if the researcher wants to test the efficacy of a sleeping pill in an experimental study, he sets the dose x and measures the corresponding hours of sleep y. It can therefore be assumed that x is error-free and y is subject to error.

On the other hand, if we want to investigate the relationship between "Education rate" and "Life expectancy" in an observational study, there is no reason to assume that one indicator is exact and the other is subject to error. Moreover, our knowledge of causal relationships between socioeconomic indicators is extremely vague and limited, and dependency relationships that may be valid in one context may not necessarily be valid in another.

Therefore, when examining covariation between indicators, variables must be treated symmetrically, each must be considered subject to error, and the regression line of the bivariate distribution must not show which change in y is caused by a particular increase in x, but which 'level' of y tends to correspond to a particular 'level' of x and vice versa (McGranahan et al. 1985).

---

[3] For the calculation of the correlation coefficient based on the L_1-norm, see Pareto (2024).

[4] In an experimental study, the researcher directly intervenes in the independent variables to determine the effect on the dependent variable or response variable, whereas in an observational study, the researcher merely observes the variables without directly intervening in them.

Below are some methods for investigating the covariation between individual indicators. In the overview, only linear models are considered and the terms x and y are used to denote the two variables involved, as in the traditional literature.

1. *Simple linear regression*

Due to the simplicity of the calculations and the familiarity of the theoretical support used, this is the most commonly used regression method. It is based on the *method of least squares*, which is referred to in the literature by the acronym OLS (*Ordinary Least Squares*).

Given n observations (x_i, y_i), $i = 1, 2, \ldots, n$, of an independent variable x and a dependent variable y, the aim is to find the values of the parameters m (slope) and p (intercept) of the linear equation:

$$y = mx + p$$

to minimize the sum of the squares of the deviations between the observed values and the estimated values of y:

$$Q(m, p) = \sum_{i=1}^{n} (y_i - mx_i - p)^2 = \min$$

This approach is based on the assumption that the data for x are exact, while the data for y are subject to errors. If we want to assume exact y_i and approximate x_i, it is possible to swap the role of x with that of y and determine a regression line that minimizes the sum of the squares of the deviations between the observed values and the estimated values of x.

This leads to the determination of two different regression lines, the regression line of y as a function of x and the regression line of x as a function of y, which intersect at the centroid[5] of the data.

Assuming that both variables are subject to error, it is possible to calculate the line that passes through the centroid and whose slope is equal to the geometric mean of the slopes of the two regression lines, both expressed with respect to y (Draper and Smith 1998). This line is called *geometric mean regression line* (GMRL) and has the following parameters:

$$\hat{m} = \pm\sqrt{\frac{\sum_{i=1}^{n} (y_i - \mathrm{M}_y)^2}{\sum_{i=1}^{n} (x_i - \mathrm{M}_x)^2}} = \pm\sqrt{\hat{m}_{y|x}\hat{m}_{x|y}}; \quad \hat{p} = \mathrm{M}_y - \hat{m}\,\mathrm{M}_x$$

where $\hat{m}_{y|x}$ and $\hat{m}_{x|y}$ are the slope of the regression line of y on x and the slope of the regression line of x on y, respectively, both referred to the plane (X, Y) and $\pm$ is the common sign of the two coefficients.

[5] The centroid is the point of coordinates $(\mathrm{M_x}, \mathrm{M_y})$.

The GMRL is a symmetric regression line. However, since it passes through the centroid, it is particularly sensitive to anomalous data and can change its slope even in the presence of a single outlier (Rousseeuw and Leroy 1987).

2. *Simple orthogonal regression*

When there is no obvious causal relationship between the variables (i.e. there is no clear distinction between the independent and the dependent variable), it may be more natural to consider the distances of each point from the regression along the normal to the line rather than the distances calculated parallel to the Cartesian axes.

This method, known in the literature under the acronym ORLS (*ORthogonal Least Squares*), consists of minimizing the sum of the squares of the perpendicular distances of the points from the regression line, i.e.:

$$Q(m, p) = \frac{\sum_{i=1}^{n} (y_i - mx_i - p)^2}{1 + m^2} = \min$$

In this way, it is assumed that both variables are subject to errors and the parameters of a single regression line are determined as follows:

$$\hat{m} = \frac{\mathrm{S}_y - \mathrm{S}_x + \sqrt{(\mathrm{S}_y - \mathrm{S}_x)^2 + 4\mathrm{S}_{xy}^2}}{2\mathrm{S}_{xy}}; \quad \hat{p} = \mathrm{M}_y - \hat{m}\,\mathrm{M}_x$$

where S_x and S_y are the standard deviation of x and the standard deviation of y, respectively, and:

$$\mathrm{S}_{xy} = \sqrt{\frac{1}{n} \sum_{i=1}^{n} (x_i - \mathrm{M}_x)(y_i - \mathrm{M}_y)}$$

The ORLS method, which is based on the least squares method, only provides an excellent description of the regression line if the variables are subject to random errors and do not contain any outliers.

3. *Orthogonal least absolute values regression*

If we replace the squares of the distances in the ORLS regression, with the absolute values, we obtain the so-called Euclidean regression, which is also referred to in the literature as ORLAV (*ORthogonal Least Absolute Values*). This regression assumes that all errors have the same importance (they are not squared) and consists of determining the regression line that minimizes the sum of the orthogonal distances of the points, i.e.:

$$Q(m, p) = \frac{\sum_{i=1}^{n} |y_i - mx_i - p|}{\sqrt{1 + m^2}} = \min$$

The calculation of the parameters of the ORLAV regression line is more complex than in the previous cases due to the absolute values and cannot be solved directly, but requires iterative procedures.[6]

However, the resulting straight line is more robust and more resistant to outliers.

4. *L_1-norm symmetric regression*

This is the median line that corresponds to the correlation coefficient based on the L_1-norm and is referred to in this context by the acronym BFL. In this case, the ratio between the sum of the absolute deviations of the points from the BFL and the sum of the absolute deviations of the points from the median line with opposite slope, called CL, is minimized:

$$Q(m) = \frac{\sum_{i=1}^{n} |y_i - mx_i - p(m)|}{\sum_{i=1}^{n} |y_i + mx_i - p(-m)|} = \min$$

where $p(m) = \underset{j}{median}(y_j - mx_j)$.

Due to the symmetry resulting from the two lines with opposite slopes, the ratio of the absolute deviations is independent of the variable for which it is calculated, so that the result does not change if the role of x is swapped with that of y.

Even in this case, an ad hoc algorithm is required to calculate the parameters of the BFL and its CL. However, once the parameters have been determined, it is also possible to obtain the absolute ratio correlation r' (Pareto 2024), unlike the ORLAV regression line, which does not have its own robust correlation measure.

The BFL is a very robust line, even in the presence of several outliers, and is particularly suitable for indicators with skewed distributions.

Figure 3.3a shows the bivariate distribution of five points, (1, 1), (2, 1.25), (3, 1.5), (4, 1.75), (5, 2), which lie along an ascending straight line. In this case, all regression lines match and fit the data perfectly. Therefore, there is a maximum positive correlation between x and y and we have: $r = \rho = r' = 1$. Let us now assume that the value of x_2 is equal to 10 (twice the maximum of x) instead of 2. Figure 3.3b illustrates this situation in which the second point has been shifted to the right. This observation is an *outlier* that has a particularly strong effect on the GMRL, ORLS and ORLAV regression lines, as it significantly changes their slope. The only line that is not affected by the anomalous value in this case is the BFL, which retains the same parameters as in the original situation. The Bravais-Pearson correlation coefficient is also highly biased and provides a completely unreliable measure ($r = 0.094$), while the Spearman rank correlation coefficient, which is based on ranks, is more robust ($\rho = 0.400$). Finally, the absolute ratio correlation provides the highest value for the degree of linear relationship between the two variables ($r' = 0.429$).

[6] For the calculation of the ORLAV line, see e.g. Bargiela and Hartley (1993).

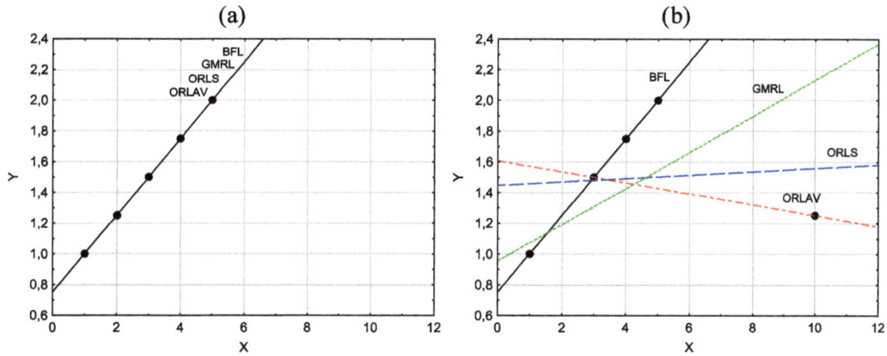

Fig. 3.3 Regression lines with an outlier

3.3.3 Multivariate Analysis

Multivariate analysis aims to examine the overall structure of the set of individual indicators in order to assess their suitability.

It is generally carried out using principal component analysis. However, in the case of pronounced nonlinear relationships, it may also be useful to perform a nonlinear principal component analysis.

Principal component analysis

Principal Component Analysis (PCA) is a statistical technique that simplifies the exploratory analysis by eliminating the redundancy that may be present in a complex set of individual indicators. This result is achieved by replacing a set of correlated variables with a smaller set of uncorrelated variables (*principal components* or *factors*) that are linear combinations of the original variables and retain most of the total variance (the first component has the highest variance, the second has an immediately lower variance, etc.).

The principal components have the following form:

$$w_{ik} = \sum_{j=1}^{m} a_{jk} x_{ij} \quad (k = 1,\ldots,m)$$

where w_{ik} is the value of component k for unit i and a_{jk} is the weight (loading) of indicator j for component k, with:

$$\sum_{j=1}^{m} a_{jk}^2 = 1 \quad (k = 1,\ldots,m)$$

and

$$\sum_{j=1}^{m} a_{jk} a_{jk'} = 0 \quad (k \neq k';\ k = 1, \ldots, m;\ k' = 1, \ldots, m)$$

The first condition is used to fix the scale of the new variables, while the second condition ensures that they are uncorrelated with each other.

The mathematical problem that arises in this procedure is therefore to find a method for determining the weights a_{jk} ($j = 1, \ldots, m$; $k = 1, \ldots, m$) so that the components have the desired properties. This is equivalent to determining the so-called eigenvalues and eigenvectors of the correlation matrix of the individual indicators. If the matrix is non-singular (i.e. there are no exact linear dependencies between the indicators), then there are m eigenvalues (latent roots) and m corresponding eigenvectors (latent vectors).

It can be shown that the largest eigenvalue, denoted as λ_1, is the variance of the first principal component and the associated eigenvector (a_{11}, a_{12}, …, a_{1m}) is the set of weights that define the first principal component. The second largest eigenvalue, denoted as λ_2, is the variance of the second principal component and the associated eigenvector (a_{21}, a_{22}, …, a_{2m}) is the set of weights that define the second principal component. The m-th or last eigenvalue, denoted as λ_m, is the variance of the last principal component and the associated eigenvector vector (a_{m1}, a_{m2}, …, a_{mm}) is the set of weights that define the last principal component (Dunteman 1989).

Furthermore, the sum of the variances is equal to m (trace of the correlation matrix[7]), i.e.:

$$\sum_{k=1}^{m} S_{w_k}^2 = \sum_{k=1}^{m} \lambda_k = m$$

with $\lambda_1 \geq \lambda_2 \geq \cdots \geq \lambda_m$.

Therefore, the proportion of the total variance explained by component k is λ_k / m, while the proportion explained by the first p components together is:

$$\frac{\sum_{k=1}^{p} \lambda_k}{m}$$

The merit of PCA is its ability to 'condense' most of the variance of the original variables into the first components. Thus, if the first p principal components, with $p \ll m$, explain a sufficient proportion of the total variance, they can be used in place of the original variables without much loss of information.

There are a number of criteria that we can use to decide how many components should be retained (stopping rules). In detail:

---

[7] The trace of a square matrix is the sum of the diagonal elements of the matrix.

1. Retain the first p components that explain a 'large' proportion of the total variance, e.g. 70–80%.
2. Retain only the components with eigenvalues greater than 1, the so-called *Kaiser rule* (Kaiser 1960).
3. Examine a *scree plot*,[8] look for the 'elbow' in the curve and retain all components just before the line flattens out (Cattell 1966).

PCA can play a very important role in the exploratory analysis of individual indicators. In particular, the method makes it possible to:

- Reduce the complexity of the data by visualizing the set of units in a small-dimensional space (e.g. a Cartesian plane). In this way, groups of units (e.g. countries, regions or provinces) with similar characteristics can be easily identified.
- Compare the empirical dimensions (factors) with the theoretical dimensions (pillars) to assess any differences or discover possible dimensions that were not taken into account when defining the phenomenon.
- Analyse the correlations between the individual indicators and the factors obtained to identify redundant and non-redundant indicators and examine the relationships between the factors and other relevant measures (e.g. "GDP per capita").

However, as already mentioned in Sect. 3.2, the use of PCA must be compatible with the chosen measurement model.

If the model is reflective (the indicators are considered as 'effects' of the phenomenon being measured and are interchangeable), the empirical dimensions should be consistent with the theoretical ones, i.e. the factors identified should represent the pillars of the theoretical framework. In this case, some indicators that correlate with each other could be redundant and omitted. Therefore, there is an overlap between the concept of correlation and that of substitutability: if two individual indicators are highly correlated, they are interchangeable (as they represent the same factor); if two indicators are poorly correlated, they are not interchangeable (as they represent different factors).

If the model is formative (the indicators are considered the 'cause' of the phenomenon being measured and are not interchangeable), the empirical dimensions simply represent groups of indicators that are correlated with each other but do not correspond to the theoretical dimensions. In this case, correlated indicators do not necessarily have to be redundant. Therefore, there is no overlap between the concept of correlation and that of substitutability. This is because correlation is an 'empirical' aspect, while substitutability is a 'theoretical' aspect that is determined by the researcher when defining the phenomenon.

---

[8] A *scree-plot* is a graph of the eigenvalues against the component number.

For example, if we want to construct a composite index of healthcare performance, the individual indicators "Number of hospital beds per 1000 inhabitants" and "Number of hospital doctors per 1000 inhabitants" are considered non-substitutable from a formative perspective, even if the two indicators are positively correlated. Conversely, in a composite index measuring leisure time use, the indicators "Proportion of people who participated in sports activities" and "Proportion of people who participated in cultural events" could be considered substitutable, even if they are not correlated.

In light of the above, PCA should only be used in formative models for exploratory purposes and not to create the composite index. The reason for this is that PCA is based on the structure of correlations between variables, but the correlations are not explained by a formative model. For an example of the use and misuse of PCA in the study of socioeconomic phenomena, see Mazziotta and Pareto (2019).

Nonlinear principal component analysis

Nonlinear Principal Component Analysis (NLPCA) is the nonlinear equivalent of PCA and allows the treatment of all types of variables (nominal, ordinal and numeric) (Casacci 2023). However, in the context of the construction of composite indices, it is applied to numeric variables (the individual indicators) in order to 'model' nonlinear relationships.

Indeed, the method performs a transformation of the original variables in order to linearize the relationships between them. This is achieved by maximizing the first p eigenvalues of the correlation matrix of the transformed variables, where p is the number of components chosen for the analysis. In other words: NLPCA transforms the original variables in such a way that the variance explained by the first p principal components of the transformed variables is maximized (Linting et al. 2007).

Thus, while in classical PCA the new variables (components) are linear combinations of the original variables, in NLPCA the new variables are linear combinations of the transformed variables. In order to 'model' nonlinear relationships between numerical variables, the transformation must be monotonic, i.e. the order between the set of original values and the set of transformed values must be maintained.

The steps for performing NLPCA are summarized below (Gifi 1990):

1. Transform each variable so that correlations between variables can increase.
2. Update the unit scores.
3. Calculate a PCA of the matrix of unit scores.
4. Repeat steps 1–3 until the sum of the p largest eigenvalues derived in step 3 is maximized.

If there are nonlinear relationships between the original variables, NLPCA will yield a variance explained by the first p components that is larger than that of the classical PCA.

A quick way to check whether nonlinear relationships exist in a set of individual indicators is therefore to compare the variance explained by NLPCA with that of classical PCA. If the two values are similar, the relationships are nearly linear; the higher the value provided by NLPCA, the greater the presence of nonlinear relationships.

An application of NLPCA in the construction of composite indices can be found at point 7 in Sect. 3.5.3.

3.4 Normalization of the Individual Indicators

Normalization aims to make individual indicators comparable, as they are often expressed in different units of measurement and/or orders of magnitude and may have different polarities.

The existence of different orders of magnitude leads to an implicit weighting of the individual indicators in proportion to their range of variation (Booysen 2002). As a result, the composite index is more strongly influenced by indicators with greater variability ('variability effect'). In addition, for individual indicators with negative polarity, it is necessary to reverse the polarity so that an increase in the normalized indicator corresponds to an increase in the composite index (Salzman 2003).

Therefore, the aim of normalization is to:

- Bring all individual indicators to the same scale or 'range', by converting them into pure, dimensionless numbers.
- Bring all individual indicators to a positive polarity.

Formally, we need to move from the original data matrix $\mathbf{X}_{n,m} = \{x_{ij}\}$ with n rows (statistical units)[9] and m columns (individual indicators) to the normalized matrix $\mathbf{Y}_{n,m} = \{y_{ij}\}$:

$$\mathbf{X}_{n,m} = \begin{pmatrix} x_{11} & \dots & x_{1j} & \dots & x_{1m} \\ \dots & \dots & \dots & \dots & \dots \\ x_{i1} & \dots & x_{ij} & \dots & x_{im} \\ \dots & \dots & \dots & \dots & \dots \\ x_{n1} & \dots & x_{nj} & \dots & x_{nm} \end{pmatrix} \Rightarrow \mathbf{Y}_{n,m} = \begin{pmatrix} y_{11} & \dots & y_{1j} & \dots & y_{1m} \\ \dots & \dots & \dots & \dots & \dots \\ y_{i1} & \dots & y_{ij} & \dots & y_{im} \\ \dots & \dots & \dots & \dots & \dots \\ y_{n1} & \dots & y_{nj} & \dots & y_{nm} \end{pmatrix}$$

where x_{ij} is the original value of indicator j for unit i and y_{ij} is the normalized value of indicator j for unit i.

There are several normalization methods, some of which bring the range or variance of the indicators to a common base and others use percentage deviations. In this text, the following cases are considered: *no normalization*, *ranking* (or rank

[9] Note that n is the number of units in the population and not in a sample.

transformation), *standardization* (or *z*-scores transformation), *Min–Max method* (or re-scaling), *indicization* (or index number transformation), *constrained Min–Max method* (or re-scaling with a reference).

The researcher must be able to determine the most appropriate normalization method for his case by taking into account the characteristics of the different methods and his needs. Since different normalization methods lead to different results, a robustness analysis must be performed in the validation step of the composite index to assess its impact on the results (Freudenberg 2003).

3.4.1 Polarity Processing

When a composite index is to be constructed, all individual indicators must have positive polarity, so that increasing each of them leads to an increase in the composite index; therefore, it is necessary to reverse the sign of the indicators with negative polarity.

The polarity reversal can be performed before normalization or during normalization itself. In most cases, however, the results match.

There are two main methods for reversing the polarity: the *linear transformation* and *the nonlinear transformation.*

1. The *linear transformation* consists of calculating the complement with respect to the maximum value of the indicator, as follows:

$$x'_{ij} = \max_i (x_{ij}) - x_{ij} \tag{3.1}$$

   where $\max_i (x_{ij})$ is the maximum of indicator j.

   This technique is the simplest and makes it possible to retain the 'distances' between the statistical units by translating the reference origin. It is usually used together with the rank transformation, standardization, and Min-Max method.
2. The *nonlinear transformation* consists of calculating the inverse of the indicator value:

$$x'_{ij} = \frac{1}{x_{ij}} \tag{3.2}$$

   This technique is often used together with the index number transformation, but since it changes the 'distances' between the statistical units and then the correlations between the indicators, it can be easily criticized (Terzi and Moroni 2004). Moreover, it can only be applied to indicators with positive values greater than zero.

Sometimes the polarity of a single indicator can be positive below a certain threshold and negative above it, or vice versa. For example, if the phenomenon to be

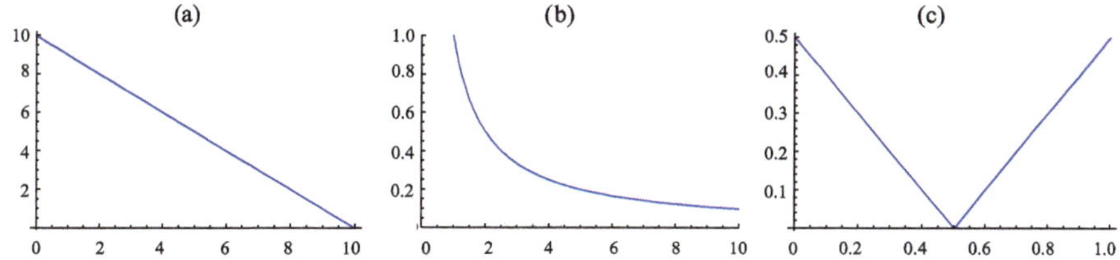

Fig. 3.4 Linear **a** nonlinear **b** and triangular **c** transformation

measured is gender equality, the indicator "Proportion of women elected to parliament out of the total number of elected representatives" has a positive polarity below 50% and a negative polarity above 50%. Similarly, if we want to measure the quality of life, the indicator "Population density (number of inhabitants per km^2)" could have a positive polarity below a certain value (for underpopulation) and a negative polarity above (for overpopulation). In this case, one speaks of a 'double polarity' of the indicator.

The simplest method to change from an indicator with double polarity to the standard case (indicator with positive or negative polarity) is the *triangular transformation.*

3. The *triangular transformation* has the form:

$$x'_{ij} = \text{abs}\,(\lambda_j - x_{ij}) \tag{3.3}$$

where λ_j is the threshold value of indicator j. If the polarity obtained is negative, a further (linear or nonlinear) transformation is required to bring the indicator to positive polarity.

Figure 3.4 shows three examples of a *linear transformation*, a *nonlinear transformation*, and a *triangular transformation.*

3.4.2 Normalization Methods

This section describes the most important methods for normalizing the individual indicators.

1. *No normalization*

The first method is to simply leave the original indicators unchanged. This solution can be chosen if all the individual indicators are expressed in the same unit of measurement and have similar ranges of variation or are expressed as a percentage. Otherwise, the lack of normalization leads to an implicit weighting of the individual indicators in proportion to their variability.

2. *Ranking* (or rank transformation)

This method consists of ranking the statistical units according to each indicator (from largest to smallest value) and replacing the value that the indicator assumes with the ordinal number with which the unit is placed in the corresponding ranking. Formally, we move from the matrix $\mathbf{X}_{n,\,m}$ to the matrix $\mathbf{Y}_{n,\,m}$ with:

$$y_{ij} = \text{rank}(x_{ij}) \tag{3.4}$$

where $\text{rank}(x_{ij})$ is the position of unit $\underline{i}$ according to indicator j.

If two or more units have the same value, they are assigned the average rank of the positions they would have occupied if they had different values, so that the sum of the ranks equals $n(n+1)/2$.

If indicator j has a negative polarity, the order must be reversed (from the lowest to the highest value). This is equivalent to applying the transformations (3.1) or (3.2) and formula (3.4).

The rank transformation is based on ordinal values and is not influenced by outliers. However, it does not take into account the 'distances' between the statistical units, so that some of the information is lost. It is therefore a rough method that makes it possible to compare the performance of units in both spatial and temporal terms only in terms of relative positions (rankings). In this case, the improvement of a position for a unit from one period to another is not necessarily linked to an improvement in the original value.

3. *Standardization* (or *z*-score transformation)

Standardization converts indicators to a common scale with zero mean and unit variance. The formula is as follows:

$$y_{ij} = \frac{x_{ij} - \mathrm{M}_{x_j}}{\mathrm{S}_{x_j}} \tag{3.5}$$

where $\mathrm{M}_{x_j} = \frac{1}{n}\sum_{i=1}^{n} x_{ij}$ and $\mathrm{S}_{x_j} = \sqrt{\frac{1}{n}\sum_{i=1}^{n}(x_{ij} - \mathrm{M}_{x_j})^2}$ are the mean and standard deviation of indicator j respectively.

If indicator j has a negative polarity, formula (3.5) must be multiplied by -1. This is equivalent to applying (3.1) and then (3.5).

The values obtained (*z*-scores) can be further adjusted to simplify the reading or to avoid negative values. For example, one can multiply each value by 10 and add 100 to obtain values that are easy to interpret (Booysen 2002).

Standardization does not bring the indicators to a common range, so that extreme values can influence the results. However, it has the advantage of centring the indicators around a common mean value and freeing them from the effect of 'variability'. Therefore, this method allows the performance of the units to be compared both spatially and temporally in terms of their relative position to the mean. In this case, the increase in the standardized value for a unit from one period to another means that the original value has increased compared to the new mean value (which may also have decreased), but this does not necessarily correspond to an increase in the original value.

4. *Min–Max method* (or re-scaling)

This method converts the indicators to a common range of [0, 1]. The formula used is as follows:

$$y_{ij} = \frac{x_{ij} - \min_i (x_{ij})}{\max_i (x_{ij}) - \min_i (x_{ij})} \tag{3.6}$$

where $\min_i (x_{ij})$ and $\max_i (x_{ij})$ are the minimum and maximum of indicator j respectively. These values define the range of each individual indicator and are called *goalposts*. In the case of time series, the goalposts are generally the minimum and maximum of indicator j in all periods under consideration or externally defined benchmarks.[10]

If indicator j has a negative polarity, the complement to 1 of formula (3.6) is calculated. This is equivalent to applying (3.1) and then (3.6).

Also in this case, the values obtained can be further adjusted to simplify the reading. For example, we can multiply each value by 1000 to obtain scores between 0 and 1000.

The Min–Max method is sensitive to outliers and has the disadvantage that it does not 'centre' the indicators around the mean or a reference value (for a possible solution, see the Constrained Min–Max method). However, it allows the indicators to be freed from the effect of 'variability' (albeit in a cruder way than z-scores).

The most interesting aspect of this method is the possibility to set the goalposts independently of the values of each individual indicator in a given period. This allows the performance of the units to be compared both spatially and temporally with respect to a common reference (the goalposts) that does not change from one period to another (as opposed to standardization, where the reference is the mean value of each period). In this case, the increase in the normalized value of a unit from one period to another corresponds to an increase in the original value.

5. *Indicization* (or index number transformation)

Indicization considers the percentage ratio between the original values and a reference value (base value) for each indicator. The index number is given by:

$$y_{ij} = \frac{x_{ij}}{x_{oj}} 100 \tag{3.7}$$

[10] Note that for new observations of the time series, the goalposts must remain unchanged in order to ensure the comparability of the normalized values. Otherwise, a new normalization of the original indicators must be carried out with new goalposts (Tarantola 2008).

where x_{oj} is the base of indicator j—generally, the maximum, the mean or an externally defined benchmark.

In this method, the base is set to 100 (but could also be set to 1 or 1000) and the unit is given a score proportional to the 'distance' from this base. A value greater than 100 means that the original value is above the base, while a value less than 100 means that the original value is below it.

By indicating with y'_{ij} the index number with the base x_{oj} (e.g. the base at time 0) and with y'_{ij} the index number with the base x_{1j} (e.g. the base at time 1), the ratio between the two sets of normalized values is constant and equal to the quantity:

$$\frac{y'_{ij}}{y''_{ij}} = \frac{\frac{x_{ij}}{x_{oj}}100}{\frac{x_{ij}}{x_{1j}}100} = \frac{x_{ij}}{x_{oj}}\frac{x_{1j}}{x_{ij}} = \frac{x_{1j}}{x_{oj}}$$

Taking the mean as the base value, formula (3.7) 'centres' the indicators around it, but does not free them from the effect of 'variability' (index numbers have the same CV of original indicators). This involves the introduction of an implicit weighting of the indicators.

If indicator j has a negative polarity, it is possible to preliminarily apply the transformation (3.2). However, the calculation of the index numbers is only recommended for indicators with positive polarity, since (3.2) changes the 'distances' between the statistical units. In addition, indicization is more sensitive to outliers than other methods, as the index numbers do not have a predefined range.

This method makes it possible to set the base independently of the values of each individual indicator in a given period in order to compare the performance of the units both spatially and temporally with respect to a common reference (the base). In this case, the increase in the index number of a unit from one period to another corresponds to an increase in the original value.

6. *Constrained Min–Max method* (or re-scaling with a reference)

The constrained Min–Max method converts indicators in a similar way to the Min–Max method, but uses a reference value (base value) that allows them to be 'centred' (Mazziotta and Pareto 2022b). The formula used is as follows:

$$y_{ij} = \frac{x_{ij} - x_{oj}}{\max\limits_{i}(x_{ij}) - \min\limits_{i}(x_{ij})} \tag{3.8}$$

where $\min\limits_{i}(x_{ij})$ and $\max\limits_{i}(x_{ij})$ are the minimum and maximum of indicator j, respectively, and x_{oj} is the base of indicator j.

In this method, the base is set to 0 and the range to 1. A positive value means that the original value is above the base, a negative value means that the original value is below it.

By indicating with y'_{ij} the normalized value with the base x_{oj} (e.g. the base at time 0) and with y''_{ij} the normalized value with the base x_{1j} (e.g. the base at time

1), with the same goalposts, the difference between the two sets of normalized values is constant and equal to the quantity:

$$y'_{ij} - y''_{ij} = \frac{x_{ij} - x_{oj}}{\max\limits_i (x_{ij}) - \min\limits_i (x_{ij})} - \frac{x_{ij} - x_{1j}}{\max\limits_i (x_{ij}) - \min\limits_i (x_{ij})} = \frac{x_{1j} - x_{oj}}{\max\limits_i (x_{ij}) - \min\limits_i (x_{ij})}$$

Taking the mean as the base value, formula (3.8) 'centres' the indicators around it and partially frees them from the effect of 'variability' by converting them to the same range (as in the Min–Max method).

If indicator j has a negative polarity, formula (3.8) must be multiplied by -1. This is equivalent to applying (3.1) and then (3.8).

Also in this case, it is possible to transform (3.8) to avoid negative values and set the base and the goalposts, regardless of the values of each individual indicator in a given period. In this way, the performances of the units can be compared both spatially and temporally with respect to a common reference (the base and the goalposts) that does not change from one period to another (for an application, see the calculation of the AMPI, at point 4 in Sect. 3.5.3).

3.4.3 Normalization Effect

Table 3.1 shows an example of normalization of an individual indicator x for five statistical units ($n = 5$). The table contains the indicator normalized with the previously described methods in the case of positive and negative polarity as well as the descriptive statistics of the normalized indicators (the most important characteristics are printed in bold).

The maximum was chosen as the base value for indicization and as the mean value for the constrained Min–Max method. In the case of negative polarity, the linear transformation was used for ranking, standardization and Min–Max method (classical and constrained), while the nonlinear transformation was used for indicization.

Note that the values normalized by ranking have a mean value that is $(n + 1)/2 = (5 + 1)/2 = 3$. In this case, the minimum is equal to 1 and the maximum is equal to n (5). However, both the minimum and the maximum could be equal to the mean (3) if all units have the same value and are therefore equal.

The z-scores have a mean of 0 and a variance of 1, so that the effect of 'variability' is eliminated. The units whose values are above the mean are given positive scores, while the units whose values are below the mean are given negative scores.

The values transformed with the Min–Max method vary between 0 and 1, and reversing the polarity does not change the standard deviation (Std), which is 0.34 in both cases. However, the mean of the normalized values for negative polarities is the complement to 1 of the mean for positive polarities, so that the coefficient of variation (CV) changes (91.9 vs. 55.1).

Table 3.1 Comparing normalization methods

Unit	Original indicator (x)	Normalized indicator (y)									
		Positive polarity (+)					Negative polarity (−)				
		Rank	z-score	Min–Max	Index number	Cons. Min–Max	Rank	z-score	Min–Max	Index number	Cons. Min–Max
1	450.0	1.0	1.81	1.00	100.0	0.63	5.0	− 1.81	0.00	11.1	− 0.63
2	200.0	2.5	0.00	0.38	44.4	0.00	3.5	0.00	0.63	25.0	0.00
3	200.0	2.5	0.00	0.38	44.4	0.00	3.5	0.00	0.63	25.0	0.00
4	100.0	4.0	− 0.73	0.13	22.2	− 0.25	2.0	0.73	0.88	50.0	0.25
5	50.0	5.0	− 1.09	0.00	11.1	− 0.38	1.0	1.09	1.00	100.0	0.38
Min	50.0	1.0	− 1.09	**0.00**	11.1	− 0.38	1.0	− 1.81	**0.00**	11.1	− 0.63
Max	450.0	5.0	1.81	**1.00**	**100.0**	0.63	5.0	1.09	**1.00**	**100.0**	0.38
Range	400.0	4.0	2.90	1.00	88.9	**1.00**	4.0	2.90	1.00	88.9	**1.00**
Mean	200.0	**3.0**	**0.00**	0.38	44.4	**0.00**	**3.0**	**0.00**	0.63	42.2	**0.00**
Std	137.8	1.4	**1.00**	0.34	30.6	0.34	1.4	**1.00**	0.34	31.5	0.34
CV %	68.9	45.9	–	91.9	**68.9**	–	45.9	–	55.1	74.6	–

The index numbers have a maximum value of 100 (base) and retain the original value of the CV (68.9), but only in the case of positive polarity. In the case of negative polarity, the 'distances' between the units have changed.

Finally, the values normalized with the constrained Min–Max method have a mean of 0 (base) and a range of 1. In this case, the minimum and maximum are not fixed, as in the classical Min–Max method, but the standard deviation is equal to that of the values normalized with the classical method (0.34).

To illustrate the effect of normalization on the distributions of the individual indicators, consider the three indicators shown in Fig. 3.5a.

The first indicator has an exponential distribution with a mean of 10 (Exp), the second indicator has a normal distribution with a mean of 100 (Nor) and the third indicator has a beta distribution with a mean of 33.3 (Bet). The indicators have different orders of magnitude as they are measurements of phenomena expressed in different units of measurement (Mazziotta and Pareto 2021).

Figure 3.5 also shows the distributions of the indicators normalized using ranking (b), standardization (c), Min–Max method (d), indicization (e) and constrained Min–Max method (f). For the last two methods, the mean value was chosen as the base.

As can be seen in Figure (b), the distributions of the indicators converted into ranks become uniform in the interval between 1 and n. This happens only in the case that all units have different values and thus ranks.

In standardization, which is shown in Figure (c), all distributions of the individual indicators are 'centred' around the origin (zero mean) and 'stretched' or 'shortened' so that they have the same variability and thus the same implicit weight (unit variance).

The Min–Max method, the results of which are shown in Figure (d), also makes the variances more homogeneous (but not equal) by converting all values in a common interval. However, the distributions of the indicators are not 'centred', which leads to the loss of a 'common' average reference value. For example, the normalized value 0.7 for the indicator Nor is a value above the mean, while for the indicator Bet it is a value below the mean.[11]

On the other hand, the conversion to index numbers in Figure (e) 'centres' all distributions around 100 (base = mean), but leaves the original variability (CV) and thus the implicit weight of each individual indicator completely unchanged.

Finally, the constrained Min–Max method 'centres' all distributions around 0 (base = mean) and brings all indicators to a unit range, limiting the effect of implicit weighting.

[11] Note that 0.5 is the mean of the range [0, 1] but not of the distributions and therefore cannot be used as a 'reference' for reading the results.

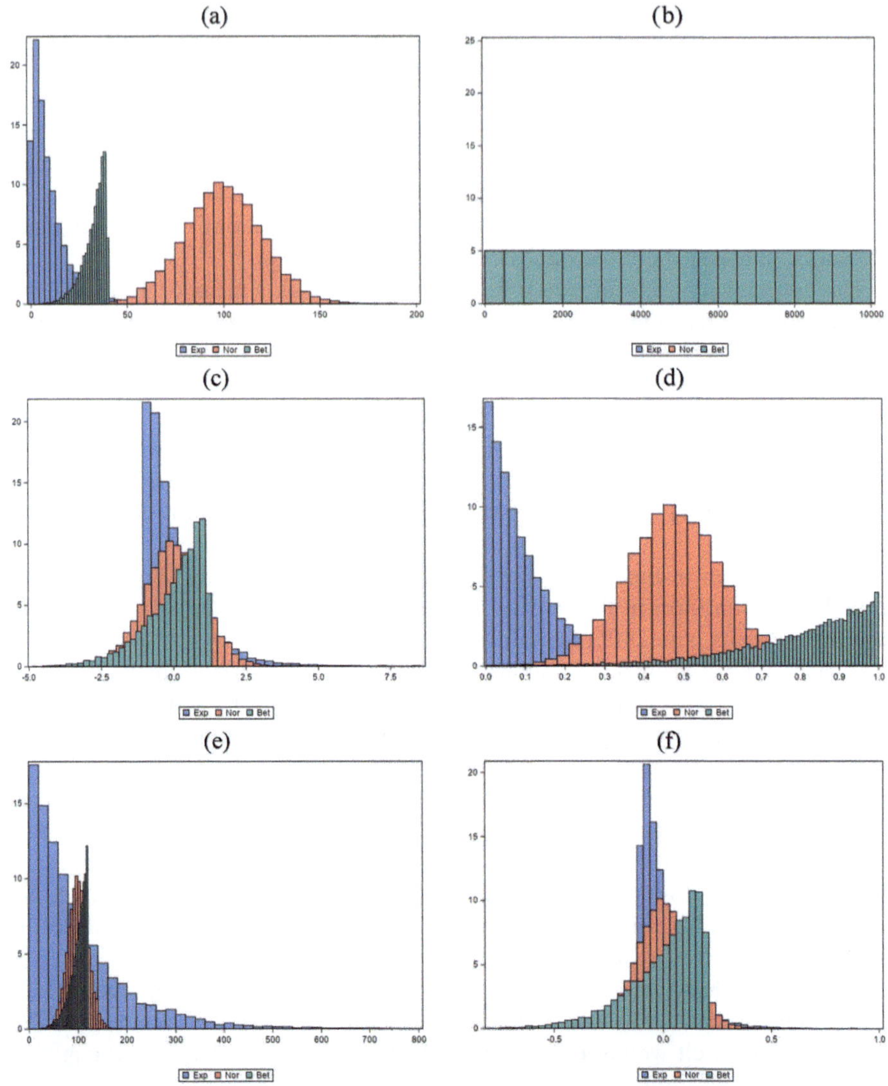

Fig. 3.5 Normalized indicators with different distributions (n = 10,000)

3.4.4 Correspondence Grid

If the values corresponding to the normalized scale for each individual indicator are given in a table, the so-called *correspondence grid* is obtained. The result is a matrix consisting of a column with the values of the common scale and as many columns as the indicators under consideration. Each row contains the values of the original indicators that correspond to each level of the normalized scale. These sets of values are considered equal or 'balanced' in the aggregation step.

The correspondence grid must be carefully constructed by the researcher as it could generate correspondences and thus balancing models that are artificial or inconsistent (Terzi et al. 2021).

Table 3.2 shows the correspondence grid of the previous indicators (Exp, Nor and Bet) for standardization, Min–Max method, indicization and constrained Min–Max method.

As we can see, all normalization methods, with the exception of the classic Min–Max method, consider the set of mean values as 'balanced'.

In standardization, a set of values is considered balanced if they are 'equidistant' from the mean in terms of standard deviations. For example, at level 3 of the correspondence grid we find all original values that are equal to the mean plus

Table 3.2 Correspondence grid by normalization method

Standardization				Min–Max method			
Scale	Exp	Nor	Bet	Scale	Exp	Nor	Bet
2.5	34.9	150.4	47.5	**1.0**	**93.3**	**186.7**	**40.0**
2.0	29.9	140.3	44.6	0.9	83.9	170.1	36.6
1.5	24.9	130.2	41.8	0.8	74.6	153.6	33.2
1.0	19.9	120.1	39.0	0.7	65.3	137.0	29.9
0.5	15.0	110.1	36.2	0.6	56.0	120.4	26.5
0.0	**10.0**	**100.0**	**33.3**	0.5	46.6	103.9	23.1
– 0.5	5.0	89.9	30.5	0.4	37.3	87.3	19.7
– 1.0	0.0	79.8	27.7	0.3	28.0	70.8	16.3
– 1.5	– 5.0	69.7	24.9	0.2	18.7	54.2	13.0
– 2.0	– 10.0	59.6	22.0	0.1	9.3	37.7	9.6
– 2.5	– 14.9	49.5	19.2	**0.0**	**0.0**	**21.1**	**6.2**
Indicization				Constrained Min–Max method			
Scale	Exp	Nor	Bet	Scale	Exp	Nor	Bet
200	20.0	200.0	66.7	0.5	56.6	182.7	50.2
180	18.0	180.0	60.0	0.4	47.3	166.2	46.9
160	16.0	160.0	53.3	0.3	37.9	149.6	43.5
140	14.0	140.0	46.7	0.2	28.6	133.1	40.1
120	12.0	120.0	40.0	0.1	19.3	116.5	36.7
100	**10.0**	**100.0**	**33.3**	**0.0**	**10.0**	**100.0**	**33.3**
80	8.0	80.0	26.7	– 0.1	0.6	83.4	30.0
60	6.0	60.0	20.0	– 0.2	– 8.7	66.9	26.6
40	4.0	40.0	13.3	– 0.3	– 18.0	50.3	23.2
20	2.0	20.0	6.7	– 0.4	– 27.3	33.7	19.8
0	0.0	0.0	0.0	– 0.5	– 36.7	17.2	16.4

three standard deviations, while at level -1 we find the values that are equal to the mean minus one standard deviation.

In the classic Min–Max method, the two sets of extreme values (including the outliers) are considered balanced and 'artificial' correspondence points (i.e. sets of artificially balanced values) are created in between. So, for example, the set of mean values corresponds approximately to the set of normalized values (0.1; 0.5; 0.8) and is considered very unbalanced. The greater the differences between the distributions of the indicators, the greater the distortion of the correspondence points.

In indicization, a set of values is considered balanced if they are 'equidistant' from the base in percentage terms. For example, at level 200 of the correspondence grid, we find all original values that are twice as large as the base (in this case the mean value), while at level 50 we find all values that are half as large. A peculiarity of this transformation is that the set of zero values of each indicator is recognized as 'balanced'. Therefore, indicization should only be used for variables that have a point of 'absolute zero' corresponding to the absence of the phenomenon or zero intensity (e.g. "Average income per capita").

The constrained Min–Max method is similar to standardization, with the difference that a set of values is considered balanced if they are 'equidistant' from the base (in this case the mean value) with respect to the range.

The advantage of indicization and constrained Min–Max method is the possibility to define a base value other than the mean value for each indicator and thus create an 'ad hoc' balancing model. This is particularly important if a non-compensatory or partially compensatory approach is chosen in the aggregation step (see Sect. 3.5.2).

3.4.5 Pros and Cons of the Normalization Methods

The main advantages and disadvantages of the various normalization methods are summarized in Table 3.3. Potential problems of normalization include loss of information (e.g. ranking), sensitivity to outliers (e.g. standardization, Min–Max method, and indicization), and implicit weighting (e.g. indicization).

Different normalization methods have different effects on the construction of the composite index and the results should be carefully evaluated (Jacobs et al. 2004). However, the method chosen should be compatible with the balancing model of the theoretical framework. For example, if a unit is considered to be balanced when it has the mean values of all indicators (or values that are proportional to it), a standardization or normalization method in which the base is set equal to the mean must be applied; while the classical Min–Max method is not correct.

Therefore, the choice of normalization method is of fundamental importance and must be carefully considered by the researcher.

Table 3.3 Pros and cons of the normalization methods

Normalization method	Pros	Cons
Ranking (or rank transformation)	Applicable to indicators with positive, negative and zero values Suitable for both bounded and unbounded indicators No/low implicit weighting (normalized indicators have equal or similar variances) Insensitive to outliers	Loss of information (from interval/ratio scale to ordinal scale) Assumes equal distances between consecutive values Aggregation by a mathematical function is questionable for ordinal data
Standardization (or z-score transformation)	Applicable to indicators with positive, negative and zero values No implicit weighting (normalized indicators have equal variances)	Not very suitable for bounded indicators Generates negative values Sensitive to outliers
Min–Max method (or re-scaling)	Applicable to indicators with positive, negative and zero values Low implicit weighting (normalized indicators have similar variances)	Not very suitable for unbounded indicators Reference to the mean may be lost Sensitive to outliers (the range depends on the extreme values)
Indicization (or index number transformation)	Suitable for both bounded and unbounded indicators Retains the coefficient of variation (only for indicators with positive polarity)	Not applicable to indicators with negative values (zero values are only accepted for indicators with positive polarity) High implicit weighting (normalized indicators have different variances) Very sensitive to outliers
Constrained Min–Max method (or re-scaling with a reference)	Applicable to indicators with positive, negative and zero values Low implicit weighting (normalized indicators have similar variances)	Not very suitable for bounded indicators Generates negative values Sensitive to outliers (the range depends on the extreme values)

* Indicators can be 'open' (unbounded) or 'closed' (bounded). An indicator is 'closed' if it has predefined limits. An example of a 'closed' indicator is the "Employment rate", which varies between 0 and 100. An indicator is 'open' if it has no predefined limits. An example of an 'open' indicator is the "Average income per capita", as there is no theoretical upper limit for income

3.5 Aggregation of the Normalized Indicators

Aggregation is the combination of all components to create one or more composite indices. This step requires defining the importance of each indicator (weighting system) and determining the technique (compensatory, partially compensatory or non-compensatory) to combine the individual indicators into a single number.

Formally, we need to move from the normalized data matrix $\mathbf{Y}_{n,m} = \{y_{ij}\}$, with n rows (statistical units) and m columns (individual indicators), to the vector $\mathbf{C}_n = \{c_i\}$, with n rows:

$$\mathbf{Y}_{n,m} = \begin{pmatrix} y_{11} & \cdots & y_{1j} & \cdots & y_{1m} \\ \cdots & \cdots & \cdots & \cdots & \cdots \\ y_{i1} & \cdots & y_{ij} & \cdots & y_{im} \\ \cdots & \cdots & \cdots & \cdots & \cdots \\ y_{n1} & \cdots & y_{nj} & \cdots & y_{nm} \end{pmatrix} \Rightarrow \mathbf{C}_n = \begin{pmatrix} c_1 \\ \cdots \\ c_i \\ \cdots \\ c_n \end{pmatrix}$$

where c_i is the value of the composite index for unit i.

A variety of aggregation methods are proposed in the literature, each of which has its advantages and disadvantages. For example, we can move from the simple arithmetic or geometric mean to multivariate statistical analysis methods. In this section, the classical methods—power mean of order r, Wroclaw taxonomic method, Principal Component Analysis—and some more recent proposals—Mean-Min function, Mazziotta-Pareto Index, performance intervals and development profiles—are presented.

Aggregation is the most important and delicate step in the construction of a composite index since the transition from the matrix $\mathbf{Y}_{n,m}$ (with $m > 1$) to the vector $\mathbf{C}_n$ always[12] involves a loss of information that must be minimized in some way. In this step, the decisions of the researcher play a fundamental role from a methodological point of view, since even a minimal change in the method used can have a strong influence on the results. Therefore, data aggregation has always been an interesting but also controversial topic in the construction of composite indicators (Saltelli 2007).

3.5.1 Weighting

Implicit weighting

As already anticipated in Sect. 3.4, normalization can lead to implicit weighting in the aggregation step if the individual indicators have not been adjusted for the different variability.

Let us suppose we have the following indicators:

[12] Unless all individual indicators are perfectly correlated so that there is redundancy.

- "Life expectancy (years)" (x_1) with a mean of 80 and a standard deviation of 1 year (range ≃ 77–83 and CV = 1.25%);
- "Hospital beds per 100,000 inhabitants" (x_2) with a mean of 1,000 and a standard deviation of 200 beds (range ≃ 400–1,600 and CV = 20%).

If we consider a significant change in x_1 from 80 to 82 years (the *z*-score increases from 0 to 2) and a non-significant change in x_2 from 1,000 to 1,025 beds (the *z*-score increases from 0 to 0.125), the percentage changes for the two indicators match and are equal to 2.5% (both index numbers increase from 100 to 102.5). Therefore, by converting the indicators into index numbers, the two changes are considered to be equivalent.

On the other hand, if we consider two changes of the same magnitude where both indicators increase by one standard deviation, i.e. x_1 increases from 80 to 81 and x_2 from 1,000 to 1,200 (both *z*-scores increase from 0 to 1), the percentage change for x_1 is 1.25% (the index number increases from 100 to 101.25), while the percentage change for x_2 is 20% (the index number increases from 100 to 120). Thus, by normalizing the indicators using indicization, the change in the indicator with the highest CV (x_2) is considered more important than the change in the indicator with the lowest CV (x_1).

Table 3.4 shows an example of the aggregation of x_1 and x_2 by converting them into *z*-scores and index numbers.

The aggregation function is a simple arithmetic mean. Consider units 1 and 5. In unit 1, x_1 is 1.5 standard deviations below the mean (78.5 years) and x_2 is 1.5 standard deviations above the mean (1,300 beds). In contrast, in unit 5, x_1 is 1.5 standard deviations above the mean (81.5 years) and x_2 is 1.5 standard deviations below the mean (700 beds).

If we want to assign equal importance to the indicators, the two units should occupy the same place in the ranking according to the arithmetic mean, as they are in a similar situation.

Table 3.4 Example of implicit weighting

Unit	Original values		*z*-scores			Index numbers		
	x_1	x_2	x_1	x_2	Mean	x_1	x_2	Mean
1	78.5	1,300	– 1.5	1.5	**0.0**	98.1	130.0	**114.1**
2	80.5	1,100	0.5	0.5	0.5	100.6	110.0	105.3
3	80.0	1,000	0.0	0.0	0.0	100.0	100.0	100.0
4	79.5	900	– 0.5	– 0.5	– 0.5	99.4	90.0	94.7
5	81.5	700	1.5	– 1.5	**0.0**	101.9	70.0	**85.9**
Mean	80.0	1,000	0.0	0.0		100.0	100.0	
Std	1.0	200	1.0	1.0		1.25	20.0	
CV %	**1.25**	**20**				1.25	20.0	

However, if we use index numbers, x_2 has a greater implicit weight than x_1 when calculating the mean, and unit 1 receives a higher score than unit 5 (114.1 versus 85.9) because it has a higher value (1,300) of x_2. If the z-scores are used instead, the two units receive the same score (0.0). This is due to the fact that standardization removes effect of 'variability'; while indicization retains the variability (i.e. the CV) of the original indicators (Mazziotta and Pareto 2012).

Explicit weighting

In the aggregation step, an explicit weighting can be defined in addition to the implicit weighting introduced during normalization.

The purpose of explicit weighting is that each weight should reflect the relative importance (relevance, significance or reliability) of the corresponding individual indicator. The explicit weights assigned to the individual indicators have a major influence on the values of the composite index. Therefore, the weights should be determined on the basis of a clearly defined theoretical framework.

The most commonly used techniques for explicit weighting of individual indicators are as follows: (1) no weighting or assignment of 'equal' weights, (2) subjective or expert weighting, and (3) weighting by PCA.

1. If no explicit weighting is defined, the individual indicators are given equal weights in addition to the implicit weighting resulting from normalization. This implies that all components of the composite index have the same importance, net of the implicit weighting, which is not necessarily correct. However, if there are no precise theoretical or empirical reasons for choosing different weights, this may be a good solution in different contexts.
2. Subjective or expert weighting is an arbitrary weighting made by the researcher or experts on the phenomenon who determine the weight of each indicator. The values obtained are then aggregated according to a specific function. Sometimes the weights are determined by policy makers or by sample surveys where respondents are asked to rate the importance of the different aspects that make up the phenomenon.
3. PCA can be used to determine the weights of the individual indicators based on the coefficients obtained for the first principal component. This empirical solution is more objective than the others and takes into account the set of weights that 'explain' most of the variance of the indicators. However, the reliability of the weights obtained depends on the variance explained by the first component and on the structure of the correlations between indicators, which does not remain constant over time. Furthermore, the weights in PCA are used to correct for overlaps between correlated indicators and are not a measure of their theoretical importance (OECD 2008). Indeed, if the individual indicators are uncorrelated, the weights cannot be estimated using this method.

Given the arbitrary nature of the various weighting criteria, and since different weighting systems lead to different results, the absence of explicit weighting (no weighting) should be the norm, while the use of a different weighting should be a theoretically justified exception (Booysen 2002).

3.5.2 Compensability Issue

A fundamental issue in the construction of a composite index is the degree of compensability or substitutability of the individual indicators.

The components of a composite index are said to be 'substitutable' if a deficit in one component can be compensated by a surplus in another component (e.g. a low value of "Percentage of people who participated in religious or spiritual activities" can be offset by a high value of "Percentage of people who participated in cultural or recreational events" and vice versa). Similarly, the components of a composite index are said to be 'non-substitutable' if no compensation between them is possible (e.g. a low value of "Life expectancy at birth" cannot be offset by a high value of "GDP per capita" and vice versa). Therefore, it is possible to define an aggregative approach as 'compensatory' or 'non-compensatory' depending on whether the substitutability of components is allowed or not (Casadio Tarabusi and Guarini 2013).

Compensability is closely related to the concept of 'imbalance', i.e. an imbalance between the indicators used to construct the composite index. In a composite index, each dimension is introduced to represent a relevant aspect of the phenomenon under consideration; therefore, measuring the imbalance between the different dimensions can contribute to the understanding of the whole phenomenon. In a non-compensatory or partially compensatory approach, all dimensions of the phenomenon must be balanced or nearly balanced. Therefore, an aggregation function must be used that implicitly or explicitly takes into account the imbalance in the form of a 'penalty'.

It is important to point out that it is up to the researcher to define 'what is meant by balanced indicators' and that the balancing model depends on the type of normalization chosen (see Sect. 3.4.4).

A compensatory approach involves the use of additive aggregation methods, such as the arithmetic mean. A non-compensatory or partially compensatory approach generally requires the use of nonlinear methods, such as the minimum or the geometric mean (OECD 2008).

3.5.3 Aggregation Methods

This section describes the most important methods for aggregating individual indicators.

1. *Power mean of order r*

The power mean of order r aggregates the normalized indicators in the following way:

$$\mathrm{M}_i^r = \left(\sum_{j=1}^{m} y_{ij}^r w_j\right)^{\frac{1}{r}}$$

where w_j is the weight of indicator j ($0 < w_j < 1$) and $\sum_{j=1}^{m} w_j = 1$.

For $r = 1$, the arithmetic mean is obtained, which is an additive aggregation function. In particular, if $w_j = 1/m$, then M_i^1 it is the simple arithmetic mean. This aggregation method has the advantage that it is methodologically 'transparent', but implies complete compensability, i.e. low values for some indicators can be compensated for by high values for other indicators.

Table 3.5 shows some special cases of the power mean of order r. The table also indicates the type of approach and the characteristics (intensity and direction) of the penalty in the case of unbalanced values. If the composite index to be constructed is 'positive', i.e. if increasing index values correspond to an improvement in the phenomenon (e.g. socioeconomic development), a downward penalty must be used (the greater the imbalance, the lower the unit score must be). On the other hand, if the composite index is 'negative', i.e. increasing values of the index correspond to a worsening of the phenomenon (e.g. poverty), an upward penalty must be applied (the greater the imbalance, the higher the unit score must be). In any case, an imbalance between the values of the individual indicators has a negative impact on the value of the composite index.

Due to the penalization (upward or downward), we have:

$$\mathrm{M}_i^{-\infty} \leq \cdots \leq \mathrm{M}_i^{-1} \leq \mathrm{M}_i^{0} \leq \mathrm{M}_i^{1} \leq \mathrm{M}_i^{2} \leq \mathrm{M}_i^{3} \leq \cdots \leq \mathrm{M}_i^{+\infty}$$

where the means are equal if and only if all values coincide, i.e. if $y_{ij} = y_{ik}$ (for each j and k with $j \neq k$).

It should be noted that not all aggregation functions are compatible with all normalization methods. For example, if individual indicators are converted into z-scores (standardization), they cannot be aggregated with the geometric mean, as this is only defined for sets of positive values.

An approach frequently used in economics is the calculation of the Jevons index (geometric mean of index numbers). This method makes it possible to construct two composite indices for each unit that are consistent with each other: a 'static' index for spatial comparisons and a 'dynamic' index for temporal comparisons (Mazziotta and Pareto 2016b).

In this perspective, the static index can be considered as the 'speed' of a unit, while the dynamic index can be considered as its 'acceleration'. Therefore, it is possible to represent the statistical units in a Cartesian diagram, the so-called *speed-acceleration diagram*, with the static index on the abscissa and the dynamic index on the ordinate or vice versa (see Sect. 3.7.2 for an example).

Table 3.5 Special cases of the power mean of order r

r	Formula	Aggregation function	Approach	Penalization	
				Intensity	Direction
$r \to -\infty$	$M_i^{-\infty} = \min_j (y_{ij})$	Minimum	Non-compensatory	Maximum	Downward
$r = -1$	$M_i^{-1} = \left(\sum_{j=1}^{m} \frac{w_j}{y_{ij}}\right)^{-1}$	Harmonic mean	Partially compensatory	High	Downward
$r \to 0$	$M_i^0 = \prod_{j=1}^{m} y_{ij}^{w_j}$	Geometric mean	Partially compensatory	Low	Downward
$r = 1$	$M_i^1 = \sum_{j=1}^{m} y_{ij} w_j$	Arithmetic mean	Compensatory	None	–
$r = 2$	$M_i^2 = \left(\sum_{j=1}^{m} y_{ij}^2 w_j\right)^{\frac{1}{2}}$	Quadratic mean	Partially compensatory	Low	Upward
$r = 3$	$M_i^3 = \left(\sum_{j=1}^{m} y_{ij}^3 w_j\right)^{\frac{1}{3}}$	Cubic mean	Partially compensatory	High	Upward
$r \to +\infty$	$M_i^{+\infty} = \max_j (y_{ij})$	Maximum	Non-compensatory	Maximum	Upward

Given a set of m individual indicators with positive polarity, let x_{ij}^t be the value of indicator j for unit i at time t, where $x_{ij}^t > 0$ ($j = 1, \ldots, m$; $i = 1, \ldots, n$; $t = t_0, t_1$). The static Jevons index can be defined as follows:

$$\mathrm{JS}_i^t = \prod_{j=1}^{m} \left(\frac{x_{ij}^t}{x_{oj}^t} 100\right)^{\frac{1}{m}}$$

where x_{oj}^t is the reference value (base) for indicator j at time t (e.g. the mean).

To compare the data between time points t_0 and t_1, the dynamic Jevons index can be calculated for each unit using the following formula:

$$\mathrm{JD}_i^{t_1/t_0} = \prod_{j=1}^{m} \left(\frac{x_{ij}^{t_1}}{x_{ij}^{t_0}} 100\right)^{\frac{1}{m}}$$

Due to the property of 'transitivity' or 'circularity' of index numbers, JS and JD are related by the relationship:

$$\mathrm{JD}_i^{t_1/t_0} = \left(\mathrm{JS}_i^{t_1} / \mathrm{JS}_i^{t_0}\right) \mathrm{JD}_o^{t_1/t_0}$$

JS and JD can only be calculated for individual indicators with positive values. Such composite indices give more weight to the low values of the individual indicators and penalize the imbalance between the components downwards.

An example of a composite index based on the power mean of order r is the United Nations Human Development Index—HDI (geometric mean of indicators normalized by the Min–Max method).

2. *Wroclaw taxonomic method*

This method was developed by a group of Polish mathematicians and applied to the aggregation of economic development indicators (Harbison et al. 1970). It is based on the concept of an 'ideal unit': a hypothetical unit that has the best value of the observed units for each indicator (optimal value).

The Euclidean distance of each statistical unit from the ideal unit is then calculated using the following formula:

$$\mathrm{d}_i = \sqrt{\sum_{j=1}^{m} (y_{ij} - y_{oj})^2}$$

where y_{ij} is the standardized value by (3.5) and y_{oj} is equal to $\min_i (y_{ij})$ or $\max_i (y_{ij})$ depending on whether indicator j has negative or positive polarity. The composite index for unit i is given by the following formula:

$$\mathrm{W}_i = \frac{\mathrm{d}_i}{\mathrm{M_d} + 2\mathrm{S_d}}$$

where $\mathrm{M_d}$ and $\mathrm{S_d}$ are the mean and standard deviation of the distances d_i respectively.

The index is equal to zero if the distance between the unit under consideration and the ideal unit is zero (all values coincide). The higher the value of the index, the greater the difference between the two units.

The main weakness of this method lies in the criterion for defining the ideal unit (Silvio-Pomenta 1973). In fact, this unit could consist of a series of values that are unattainable or inconsistent in practice. Of course, the ideal unit can also be 'constructed' according to standards that are considered optimal or that represent political goals. However, the ideal unit can vary over time, which makes comparisons over time difficult.

3. *Mean-Min Function*

The Mean-Min Function (MMF) is a two-parameter function that includes the two extreme cases of penalization of the imbalance: the zero penalty, represented by the arithmetic mean (compensatory approach), and the maximum penalty, represented by the 'minimum' function (non-compensatory approach).

The function penalizes the imbalance downwards and allows to vary the intensity of the penalty (from maximum to zero) by changing the parameters. The composite index is defined as:

$$\mathrm{MMF}_i = \mathrm{M}_{y_i} - \alpha\left(\sqrt{(\mathrm{M}_{y_i} - \min_j\{\, y_{ij}\}\,)^2 + \beta^2} - \beta\right) \quad (0 \le \alpha \le 1;\, \beta \ge 0)$$

where M_{y_i} is the mean of the normalized values for unit i and $\min_j\{\, y_{ij}\}$ is the minimum of the normalized values for unit i. The parameters α and β are a measure of the intensity of penalization of the imbalance and a measure of the degree of complementarity of the individual indicators (Casadio Tarabusi and Guarini 2013).

The function reduces to the arithmetic mean for $\alpha = 0$ (in this case, β is irrelevant) and to the minimum for $\alpha = 1$ and $\beta = 0$. Therefore, the definition interval of the composite index values is: $\min_j\{\, y_{ij}\} \le \mathrm{MMF}_i \le \mathrm{M}_{y_i}$.

The MMF is independent of the choice of normalization method. This means that the normalized matrix $\mathbf{Y}_{n,m}$ can be obtained by standardization, re-scaling or indicization.

An appropriate choice of parameter values should yield the aggregation function that best fits the chosen theoretical approach. However, there is no general rule for determining such values (Mazziotta and Pareto 2015).

4. *Mazziotta-Pareto Index*

The Mazziotta-Pareto Index (MPI) is a composite index for the aggregation of a set of indicators that are considered not (fully) substitutable. The index is based on a nonlinear function that, starting from the arithmetic mean of the normalized indicators, introduces a penalty for the units that show an imbalance in the values of the individual indicators (De Muro et al. 2011).

There are two versions of the index: (a) MPI and (b) Adjusted MPI (AMPI). The first version of the index is the best solution for a 'static' analysis (e.g. the analysis of a single year), while the second version is the best solution for a 'dynamic' analysis (e.g. the analysis of time series).

(a) *MPI*

The MPI is based on the following normalization:

$$z_{ij} = 100 + 10y_{ij}$$

where y_{ij} is given by (3.5). Therefore, the normalized indicators have a mean of 100 and a standard deviation of 10.

If we use M_{z_i} and S_{z_i} to denote the mean and standard deviation of the normalized values of unit i respectively, and with $\mathrm{cv}_{z_i} = \mathrm{S}_{z_i}/\mathrm{M}_{z_i}$ the corresponding coefficient of variation, the composite index is defined as follows:

$$\mathrm{MPI}_i^{+/-} = \mathrm{M}_{z_i} \pm \mathrm{S}_{z_i}\mathrm{cv}_{z_i}$$

where the sign ± depends on the type of phenomenon being measured. To apply a downward penalty, the MPI^- must be used; to apply an upward penalty, the MPI^+ must be used.

The MPI thus decomposes the score of each unit into two parts: the mean level (M_{z_i}) and the penalty ($\mathrm{S}_{z_i}\mathrm{cv}_{z_i}$). The penalty is a function of the variability of the indicators with respect to the unit's mean ('horizontal variability'), and is used to penalize the unit if it is unbalanced. The aim is to reward the units that have a greater balance among the values of the individual indicators with the same mean value.

(b) *AMPI*

The AMPI normalizes the indicators in the following way:

$$r_{ij} = 100 + 60y_{ij}$$

where y_{ij} is given by (3.8). Therefore, the normalized indicators have a base of 100 and a range of 60.

If we use M_{r_i} and S_{r_i} to denote the mean and standard deviation of the normalized values of unit i, and with $\mathrm{cv}_{r_i} = \mathrm{S}_{r_i}/\mathrm{M}_{r_i}$ the corresponding coefficient of variation, the composite index is defined as follows:

$$\mathrm{AMPI}_i^{+/-} = \mathrm{M}_{r_i} \pm \mathrm{S}_{r_i}\mathrm{cv}_{r_i}$$

where the sign ± depends on the type of phenomenon being measured. To apply a downward penalty, the AMPI^- must be used; to apply an upward penalty, the AMPI^+ must be used.

The main difference between MPI and AMPI lies in the normalization method. The MPI is based on a standardization of the individual indicators, which is repeated independently in each period under consideration, so that it is not possible to assess 'absolute' changes in the values of the units over time. The AMPI is based on the constrained Min–Max method and makes it possible to assess the 'absolute' changes in the values of the units with respect to a base and pre-set goalposts that remain constant over time.[13]

In addition, the AMPI allows the score of each unit to be calculated independently of the others, while the MPI requires knowledge of the mean and standard deviation of each indicator.[14]

[13] To obtain normalized values that are 'centred' around 100, the base of indicator j should be a 'typical' value that is as close as possible to the central value.

[14] For a detailed comparison between the two versions, see Mazziotta and Pareto (2016a).

5. *Principal components*

We have seen that PCA is a statistical technique of multivariate analysis that allows starting from a large number of individual indicators, to construct a limited number of composite indices (principal components or factors) that 'explain' most of the observed variance. The first principal component is often used as the 'best' composite index and is defined as follows:

$$C_{i1} = \sum_{j=1}^{m} a_{j1} x_{ij} \tag{3.9}$$

where a_{j1} is the weight (loading) of indicator j for factor 1 (first component).

The resulting index has special mathematical properties, as it 'explains' most of the variance of the individual indicators. This is achieved by maximizing the sum of the squares of the correlation coefficients between the composite index (first principal component) and the individual indicators.

However, the first principal component only explains a limited part of the variance, so that a considerable amount of information may be lost.

Moreover, a composite index based on PCA is often 'elitist' (Mishra 2007), as it has a strong tendency to represent highly correlated indicators and neglect the others, regardless of their conceptual importance. Therefore, important individual indicators that are only slightly correlated with each other cannot be correctly represented by the composite index.

Finally, the signs of the weights in (3.9) are based on the observed correlations and not on the polarities defined by the user, as is required in a formative model. In fact, even if the individual indicators are all brought to positive polarity, during the normalization step (e.g. by changing the sign of the z-scores), the PCA-based composite index ignores the polarities and provides consistent results if and only if:

$$\text{sgn}(r_{x_j,x_k}) = p_j p_k \quad \text{for } j \neq k \tag{3.10}$$

where p_j and p_k are the polarities of indicators j and k.

Equation (3.10) is the formalization of point 5 of Table 2.1 for reflective models. If a formative model is assumed and (3.10) is not satisfied, the composite index will give inconsistent results[15] (see Sect. 4.2.1). For these reasons, the first principal component should only be used as a composite index in a reflective approach.

An alternative method for constructing a PCA-based composite index is the weighted average of the factors (Giudici and Avrini 2002). In this approach, the individual indicators are aggregated by a weighted mean of the factor scores, with

[15] For a numerical example, see Mazziotta and Pareto (2024).

weights proportional to the variance explained by the factors. The composite index for unit i is:

$$S_i = \frac{\sum_{h=1}^{p} C_{ih}\lambda_h}{\sum_{h=1}^{p} \lambda_h}$$

where C_{ih} is the value of factor h for unit i, λ_h is the percentage of variance explained by factor h, and p is the number of factors considered ($p \leq m$). If $p = m$, there is no loss of information. In this method, the factors (from the first to the last) are assigned decreasing importance according to their proportion of variance explained.

6. *Performance interval*

An alternative approach for constructing a composite index is to calculate a range of possible values for each statistical unit instead of a single value. This range is called 'performance interval' and can be formed by varying the degree of compensability of the individual indicators (Mazziotta and Pareto 2022a).

The performance interval defines a lower and an upper bound for the composite index: one bound corresponds to the hypothesis of maximum (or full) compensability of the individual indicators (compensatory composite index), while the other corresponds to the hypothesis of non-compensability of the individual indicators (non-compensatory composite index). From this point of view, the middle of the range can be considered as the value of the composite index in the partial compensability hypothesis (partially compensatory composite index).

Based on Table 3.5, it is possible to define two different performance intervals, depending on the type of phenomenon to be measured.

Given the normalized matrix $\mathbf{Y}_{n,m} = \{y_{ij}\}$, we have:

(a) '*Positive*' performance interval

$$\left(\min_j (y_{ij}); \quad \frac{1}{m}\sum_{j=1}^{m} y_{ij}w_j \right)$$

where the upper bound corresponds to the hypothesis of maximum compensability (arithmetic mean) and the lower bound corresponds to the hypothesis of non-compensability (minimum).

(b) '*Negative*' performance interval

$$\left(\frac{1}{m}\sum_{j=1}^{m} y_{ij}w_j; \quad \max_j (y_{ij}) \right)$$

where the lower bound corresponds to the hypothesis of maximum compensability (arithmetic mean) and the upper bound corresponds to the hypothesis of non-compensability (maximum).

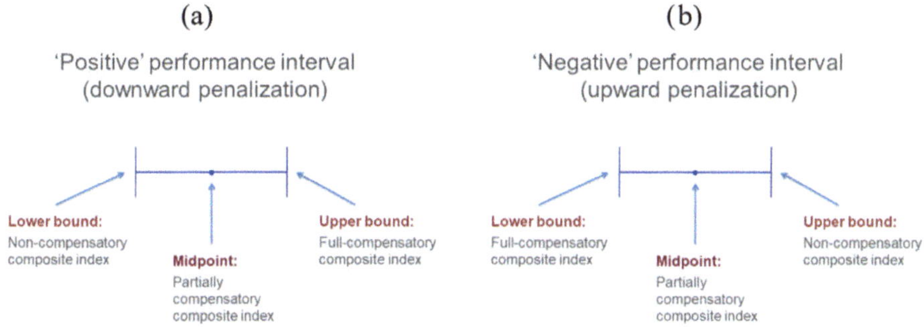

Fig. 3.6 Performance intervals

To apply a downward penalty, the 'positive' performance interval must be used; to apply an upward penalty, the 'negative' performance interval must be used. Figure 3.6 shows the two possible performance intervals.

Note that the length of the performance interval can be seen as a measure of the imbalance of the normalized indicators. If the normalized indicators are perfectly balanced, the performance interval is reduced to a single point and the value of the composite index is independent of the assumptions about the degree of compensability of the individual indicators. The longer the interval (i.e. the greater the imbalance of the normalized indicators), the larger the difference between the value of the compensatory composite index and the value of the non-compensatory composite index.

The matrix $\mathbf{Y}_{n,m} = \{y_{ij}\}$ can be obtained by applying any normalization method from Sect. 3.4.2.

7. *Development Profile*

The development profile method differs from the previous methods in that it uses a multivariate normalization technique that makes it possible to take into account nonlinear relationships between the individual indicators.

The basic idea goes back to Donald McGranahan, the founder and then Director of the United Nations Research Institute for Social Development (UNRISD) from 1967 to 1977, and was proposed as part of a study on the measurement and analysis of socioeconomic development (McGranahan et al. 1985).

The original proposal was to interpolate the data in multidimensional space with a piecewise linear function that allowed the curvilinear trend of points from the lowest to the highest level of development to be described. The indicators were then transformed in an iterative process so that the piecewise linear function was linearized and a common scale was formed. Once the common scale was found, it was possible to create a correspondence grid and draw the so-called development profiles by indicating the values of each country on the scales of the individual indicators and connecting these points with a dashed horizontal line.

Nowadays, this problem can be easily solved with NLPCA. This method makes it possible to transform the set of individual indicators in such a way that the relationships between them are linearized. The first principal component of the transformed indicators is the best interpolation line of the point cloud and its projection into the space of the original indicators provides a set of points (the correspondences between individual indicators) that allow to construct the correspondence grid. It is therefore possible to calculate a composite index for each unit by aggregating the normalized values (e.g. by the arithmetic mean) and drawing a development profile on the correspondence grid (for an example, see Sect. 4.2.2).

The procedure comprises the following steps (Casacci and Pareto 2021):

1. Given the matrix $\mathbf{X} = \{x_{ij}\}$, compute the transformed matrix $\mathbf{Y} = \{y_{ij}\}$ with $y_{ij} = f_j(x_{ij})$, where f_j is a monotonically increasing spline transformation such that the largest eigenvalue of the PCA of the matrix $\mathbf{Y}$ is maximized. This step is achieved by NLPCA of the matrix $\mathbf{X}$.
2. Given the factor loading a_{j1} of indicator j for the first factor (the BFL), calculate the predicted transformed value of indicator j using the formula $y_{zj} = z\, a_{j1}$, where z ranges from approximately -3 to $+3$ with a step of 0.1 (or as desired).
3. Construct the correspondence grid $\mathbf{G} = \{g_{zj}\}$ with $g_{zj} = f_j^{-1}(y_{zj})$, where g_{zj} is the value of indicator j corresponding to the score z of the common scale.[16]
4. Calculate the normalized matrix $\mathbf{Z} = \{z_{ij}\}$ with $z_{ij} = y_{ij}/a_{j1}$, where z_{ij} is the normalized value of indicator j for unit i.
5. Construct a development profile and/or composite index for each unit based on the normalized values.

3.6 Validation of the Composite Index

The validation step aims to assess the robustness of the composite index in terms of its ability to produce correct and stable measures, as well as, its discriminatory power. As we have seen, the results and the ranking of statistical units according to a composite index may largely depend on the choices made in the previous steps (selection of the individual indicators, normalization and aggregation). For this reason, statistical analyses must be performed to examine the robustness of the ranking when an individual indicator is included or excluded or when the methods used to construct the composite index vary (Freudenberg 2003).

The robustness of a composite index is assessed using two different methods: Uncertainty Analysis (UA) and Sensitivity Analysis (SA). UA focuses on how the

[16] If the data were normalized by z-scores, we would have the mean of each indicator at level 0 of the correspondence grid (see Sect. 3.4.4). However, this correspondence is only correct if the individual indicators are linearly interrelated.

uncertainty in the input factors is transmitted through the structure of the composite index and affects the results. SA analyses how each source of uncertainty contributes to the variance of the results (Saisana et al. 2005).

UA and SA can be used synergistically and iteratively in the construction of a composite index to facilitate the selection of individual indicators, make the process of index construction more transparent, and examine the robustness of the ranking, including with respect to possible alternative composite indices (USAID 2014).

The discriminatory power of a composite index is assessed by examining its ability to: (a) distinguish between units and/or groups of statistical units; (b) distribute all units in a 'continuum' without concentrations in some segments; (c) show values that are interpretable in terms of discriminatory power by identifying certain cut-off values (thresholds) or cut-points (Maggino and Zumbo 2012).

3.6.1 Uncertainty Analysis

UA is essentially based on simulations performed using the various equations that form the underlying model. A valid approach to assess the uncertainty of the results is the Monte Carlo method, which is based on repeated evaluations of the model by varying a set of randomly selected input factors (OECD 2008).

The steps of the procedure are as follows:

1. Identification of k input factors Fi ($i = 1, \ldots, k$), which can introduce uncertainty into the results (e.g. errors in individual indicators, exclusion of an individual indicator, etc.).
2. Assignment of a probability distribution (or density function) to each input factor (e.g. a normal distribution for the errors of the individual indicators; a discrete uniform distribution for the selection of individual indicator to be excluded, etc.).
3. Random generation of L independent combinations or samples (replicacations) of the input factors $F_1^l, F_2^l, \ldots, F_k^l (l = 1, 2, \ldots, L)$ and calculation of the corresponding value of the composite index for each unit $c_i^l = f(x_{i1}, x_{i2}, \ldots, x_{im}; F_1^l, F_2^l, \ldots, F_k^l)$ $(i = 1, 2, \ldots, n; l = 1, 2, \ldots, L)$.
4. Calculation of the average shift (displacement) of the units in the ranking [17] $\overline{\mathrm{R}}^l = \frac{1}{n}\sum_{i=1} \left|r_i^l - r_i\right| (l = 1, 2, \ldots, L)$, where $r_i^l = \mathrm{rank}(c_i^l)$ is the rank of unit i according to the composite index for the sample l and r_i is the original rank of unit i according to the composite index.
5. Analysing the distribution of $\overline{\mathrm{R}}^l$ and/or $r_i^l (l = 1, 2, \ldots, L)$. The main characteristics of these distributions, such as mean and variance, are estimated with a precision that depends on the number of samples L generated in the simulation. In general, the lower the variance, the greater the robustness.

[17] This quantity is referred to as the *mean absolute difference of rank*.

A special and very important case of UA is Influence Analysis (IA), which has the purpose of empirically quantifying the 'weight' of each individual indicator in the calculation of the composite index. Given m individual indicators, IA consists of performing steps 3 and 4 with $L = m$, excluding the individual indicator l each time. The value of $\overline{\mathrm{R}}^{l}$ represents the 'weight' of indicator l (Mazziotta and Pareto 2017).

Regarding the application of UA, two aspects should be emphasized.

The first aspect concerns the choice of input factors. Some authors suggest using the largest possible number of input factors, regardless of the theoretical framework, and investigating how the rankings change depending on the composite index when the methods of normalization, weighting, and aggregation vary (OECD 2008; Luzzati and Gucciardi 2013). However, it is important to note that only factors that are consistent with the basic assumptions should be used. For example, if it is assumed that the individual indicators are substitutable (compensatory approach), only additive aggregation functions (such as the sum or the arithmetic mean) should be tested, but not multiplicative (such as the geometric mean) or nonlinear functions. Conversely, in a non-compensatory or partially compensatory approach, only functions that—implicitly or explicitly—penalize the imbalance among the normalized indicators (such as the geometric mean or the MPI) should be considered. The same applies to normalization and weighting. If the individual indicators are assumed to have the same importance, only normalization methods that remove the variability effect (such as standardization or the Min–Max method) should be tested, and not methods that preserve the original variability (such as indicization). In addition, no different weights should be assigned to the individual indicators in the simulations.

The second aspect concerns the use of these simulations to create the composite index. Some authors suggest not to rank the units based on a single composite index, but to create many rankings based on different composite indices, calculate the frequency distribution of the positions reached by each unit and derive a 'plausible' ranking by summarizing these distributions, e.g. by the median (Luzzati and Gucciardi 2015). This approach is only valid for creating composite indices that are compatible with the theoretical framework and not, as mentioned above, with all possible methods. It also has some particularly relevant shortcomings. In particular: (a) it only allows the creation of a ranking of units without taking into account the 'distances' between them; (b) it does not allow the calculation of scores based on a well-defined model; (c) it does not provide a transparent result that can be replicated by the end user.

3.6.2 Sensitivity Analysis

SA examines the degree of influence of each input factor on the composite index and shows how much each source of uncertainty contributes to the variance of the results (OECD 2008).

The importance of a given input factor F_i can be measured by the Sensitivity Index, which is defined as the proportion of variance in the results due to the uncertainty in F_i. For k independent input factors, the sensitivity indices can be calculated using the following formula for the decomposition of the total variance of the results ($\overline{\mathrm{R}}^l$ or r_i^l):

$$\mathrm{V} = \sum_i \mathrm{V}_i + \sum_i \sum_{j>i} \mathrm{V}_{ij} + \cdots + \mathrm{V}_{12\ldots k}$$

where V_i is the variance due to uncertainty in F_i, V_{ij} is the variance due to uncertainty in the interaction between F_i and F_j, and so on.

A first measure of the proportion of variance in the results V that is due to the uncertainty in F_i is the first-order Sensitivity Index for the factor F_i, defined as:

$$\mathrm{S}_i = \frac{\mathrm{V}_i}{\mathrm{V}}$$

A measure that summarizes all interactions involving a given factor F_i in a single term is the Total Effect Sensitivity Index, which is given by the formula:

$$\mathrm{S}_{Ti} = \frac{\mathrm{V}_i}{\mathrm{V}} + \sum_{j \neq i} \frac{\mathrm{V}_{ij}}{\mathrm{V}} + \cdots + \frac{\mathrm{V}_{12\ldots k}}{\mathrm{V}} = \mathrm{S}_i + \sum_{j \neq i} \mathrm{S}_{ij} + \cdots + \mathrm{S}_{12\ldots k}$$

where S_{ij} is the second-order Sensitivity Index for factors F_i and F_j, and so on.

If the chosen model does not include for interactions between the input factors (additive model), we have $\mathrm{S}_{Ti} = \mathrm{S}_i (i = 1, 2, \ldots, k)$ and $\sum_i \mathrm{S}_{Ti} = 1$.

In general, $\sum_i \mathrm{S}_{Ti} \geq 1$ and the presence of a significant difference between S_{Ti} and S_i indicates an important role of the interaction between the factors F_i in the production of the output.

The estimators of (S_i, S_{Ti}) can be obtained using different methods, the most important of which is the method of Sobol (1993).

3.7 Presentation of the Results

Due attention should always be paid to the presentation of results, as visualization can influence (or better still improve) the interpretability of the phenomenon being measured (OECD 2008).

In order for the calculation procedure to be clear and transparent, it is necessary to present one or more tables containing both the individual indicators (input) and the composite index (output), to describe in detail the formulas used for normalization and aggregation, and to explain the reasons for this choice. The polarity and any reference values of each indicator (base or goalposts) must also be specified in order to ensure the reproducibility of the procedure.

To facilitate the understanding of the results, it is advisable to present both the composite index and the individual indicators graphically, i.e. the so-called 'dashboard' (or any pillars if a composite index is formed for each dimension), in order to be able to assess the extent of the individual components of the phenomenon under investigation.

3.7.1 Graphical Representation of the 'Dashboard'

Often, end users are not sufficiently familiar with multidimensional graphical representations, so the researcher should avoid using complex tools that may be incomprehensible. The most commonly used charts to visualize a set of individual indicators are *icon plots*, which allow even non-experts to approach statistical information.

Icon plots make it possible to display the values of some variables on statistical units by means of schematic images. The dimensions and shape of these images depend on the values assumed by the variables studied, and the regularities highlighted by the diagrams can be used to discover similarities and homogeneous groups. The graphical tools used in this technique are usually objects that can be perceived and even intuitively interpreted through simple association processes, such as radar (*radar charts*) or stars (*star charts*) and human faces (*Chernoff faces*).

Radar chart or Kiviat diagram
This type of representation uses a polar coordinate system consisting of a sequence of rays forming equal angles to each other. Each ray represents an individual indicator and a point is placed on it whose distance from the centre is proportional to the value of the indicator in relation to the maximum. The dots on the rays are then connected with a dashed line to highlight the configuration of each unit.

Since the rays have a common scale, the indicators must be normalized before graphing so that the values obtained by the transformation fall within a given interval. The most commonly used methods are Min–Max and standardization: the first method allows the entire range of indicators to be displayed, but not the mean value, and is sensitive to outliers; the second method allows the mean value to be displayed, but the extreme values must be truncated.

Figure 3.7a shows an example of a radar chart for eight indicators (X1-X8) normalized using the Min–Max method. In this case, each ray is labelled with a unit amplitude scale from 0 to 1.

The figure shows the dashed lines corresponding to the following units: the Max unit (orange colour) has the maximum value for each indicator and the Min unit (black colour) has the minimum value for each indicator (the dashed line is reduced to a dot in the centre).

In standardization the values outside the interval $(-3; +3)$ are generally approximated by the boundaries of the interval, and each ray can be given an amplitude scale of 6, from -3 to $+3$. If a value is equal to the average value of the individual indicator, the point representing it is placed in the centre of the

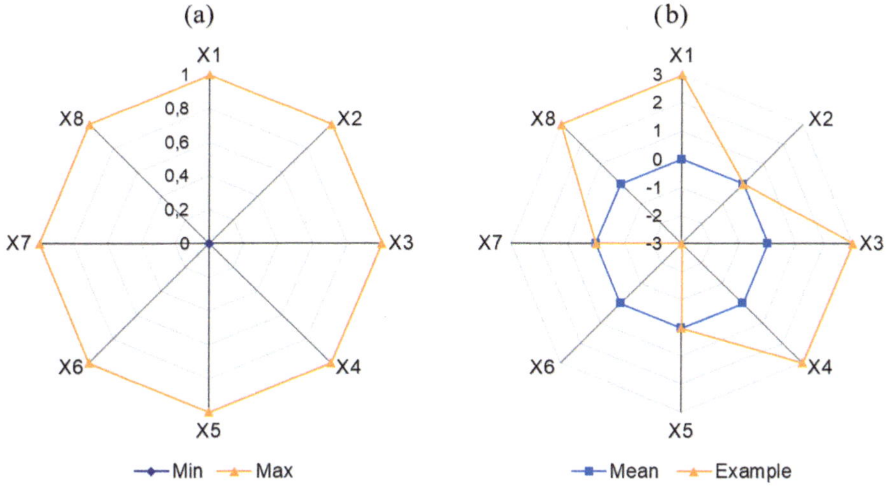

Fig. 3.7 Comparing radar charts with different data normalization

radius associated with the variable, while all values below − 3 and above + 3 can be placed at the beginning and end of the radius, respectively.[18]

Figure 3.7b shows an example of a radar chart in the case of standardization. The figure shows the dashed lines corresponding to the following units: the Mean unit (blue colour) has the mean value for each indicator, and the Example unit (orange color) has a value greater than or equal to + 3 times the standard deviation for X1, X3, X4 and X8, the mean value for X2, X5 and X7, and a value less than or equal to -3 times the standard deviation for X6.

In the radar or star chart, the assignment of the individual indicators to the rays is arbitrary, so that the shape of the dashed lines can change depending on the assignment and the area within these lines has no meaning.

Particular attention must also be paid to the minimum values of the individual indicators so that the assignment of the indicators to the rays does not lead to confusion. For example, it may happen that a unit has several minimum values that correspond to alternating rays, and in this case the dashed line coincides with the rays. However, this problem is negligible and radar charts provide a good representation of multidimensional phenomena.

In general, the phenomenon is homogeneous if the outline of the dashed line of a unit is regular and all the quantities studied have approximately the same level; if, on the other hand, it is jagged, the phenomenon is not uniform and only some quantities have a satisfactory level.

[18] According to the Bienaymé-Chebicev theorem, the frequency of cases that lie outside the interval is always less than or equal to 11%. In a normal distribution, this frequency is 0.3%.

The radar chart provides a synthetic representation of the statistical units, which makes it possible to immediately evaluate the intensity of the phenomenon and the possible relationships between different indicators (for example, if in most units the first two rays are long while the last one is short, there is a positive correlation between the first two indicators and a negative correlation with the last one). This makes the radar chart particular sensitive to simultaneous changes in the indicators, which are expressed in the polar coordinate system in clear changes in the shape of the dashed lines. Due to these properties, radar charts are particularly suitable for visualizing the structural changes of a series of indicators in space and time.

Figure 3.8 shows the radar chart of eight individual indicators for the world regions used by the UNDP (United Nations Development Programme). The variables are assigned to the rays in a clockwise direction. Looking for possible regularities, the similarities between Europe and Central Asia and Latin America/ Caribbean become evident.

Chernoff faces
Chernoff icons use our natural sensitivity to perceive and interpret facial expressions to qualitatively express the information provided by a set of quantitative variables. In this case, the variables studied are associated with the shape and size of the somatic features of a face, which are reproduced schematically, and the comparison between the images obtained allows us to identify regularities, correlations and interactions between the variables. When measuring phenomena such

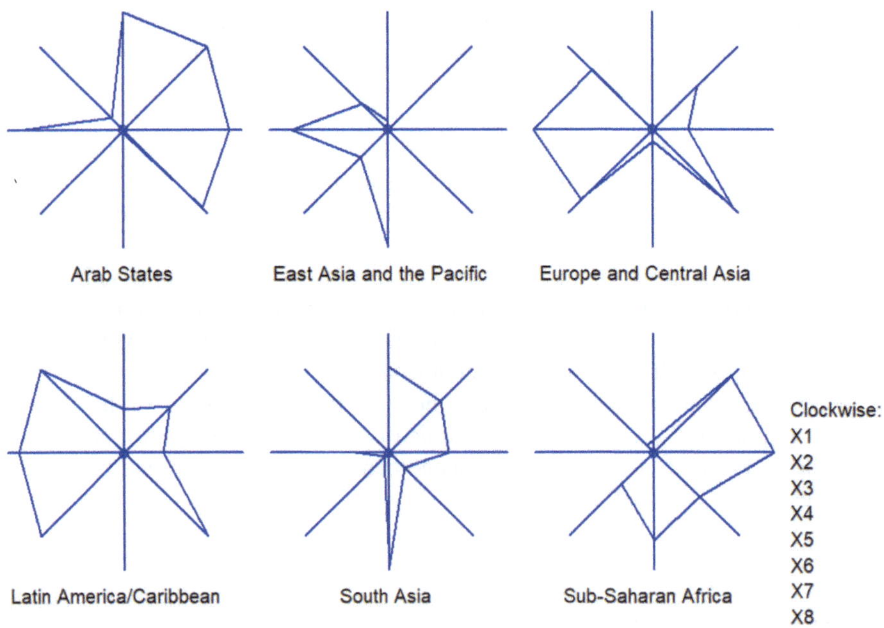

Fig. 3.8 Radar chart

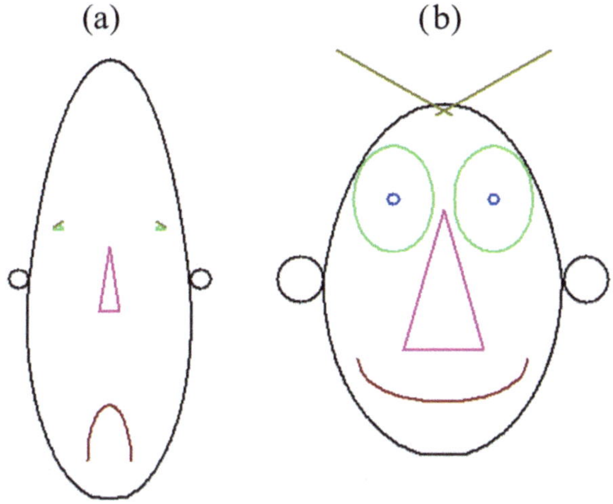

Fig. 3.9 Extreme configurations of Chernoff faces

as development or well-being, the effectiveness of the representations is further enhanced by a clear correspondence between the shape of the graphic support and the content of the information. Indeed, it is possible to associate high values of development or well-being indicators with happy and satisfied faces and low values of indicators with sad and suffering faces.

Figure 3.9 shows two examples of extreme configurations. The negative extreme (a) is associated with the ideal unit that has the worst values of all indicators, and the positive extreme (b) with the unit that has the best values. In general, the following somatic features are considered: width of face (narrow-wide), length of face (long-short), size of ears (small-large), eccentricity of eyes (oval-round), size of eyes (small-large), slope of eyebrows (high-low), length of eyebrows (short-long), width of nose (narrow-wide), height of nose (long-short), curvature of mouth (convex-concave), width of mouth (narrow-wide).

The images obtained can be compared with each other to identify regularities or homogeneous groups among the different configurations. This technique facilitates the identification of similarities, even when the differences are not very great, since it is based on the idea that people recognize faces and perceive small changes without difficulty.

As for the operational phase, keep in mind that Chernoff faces can be very different depending on the assignments made and therefore require a careful study of the best way to match variables to somatic features. Furthermore, certain features (e.g. the width of the face) are more noticeable than others, giving more importance to the variables represented by these more prominent features (Spencer 2014).

Figure 3.10 shows the Chernoff faces of the same eight individual indicators from Fig. 3.7 for the world regions. Again, the similarity between Europe and Central Asia and Latin America/Caribbean can be seen in comparison to the other regions (both faces have a similar shape, a long nose and a smiling mouth).

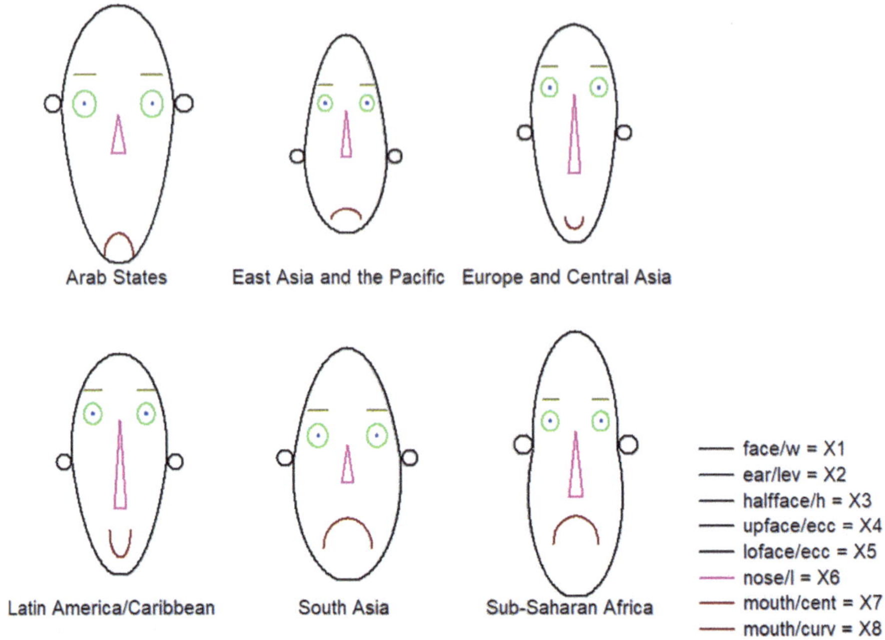

Fig. 3.10 Chernoff faces

3.7.2 Graphical Representation of the 'Synthesis'

The graphical representation of the composite index is much simpler than that of the dashboard, since we are moving from a multidimensional space to a unidimensional space.

Generally, bar charts or cartograms are used for 'static' analyses (e.g. analysing a single year) and line charts or heatmaps are used for 'dynamic' analyses (e.g. analysing time series). To visualize the relationship between the composite index and another relevant variable, the classic scatterplot can be used. Some examples are listed below.

Bar chart
With the bar chart (horizontal or vertical) we can simultaneously visualize the composite index and the corresponding ranking. For this purpose, it is sufficient to order the units from the highest to the lowest value of the composite index and plot them on the *x*-axis of a Cartesian diagram in the case of a vertical bar chart or on the *y*-axis in the case of a horizontal bar chart.

Figure 3.11 shows an example of vertical bar chart of a composite index for a set of countries with a base of 100. The vertical bar chart is particularly suitable if the number of statistical units is limited, otherwise it is better to use the horizontal bar chart.

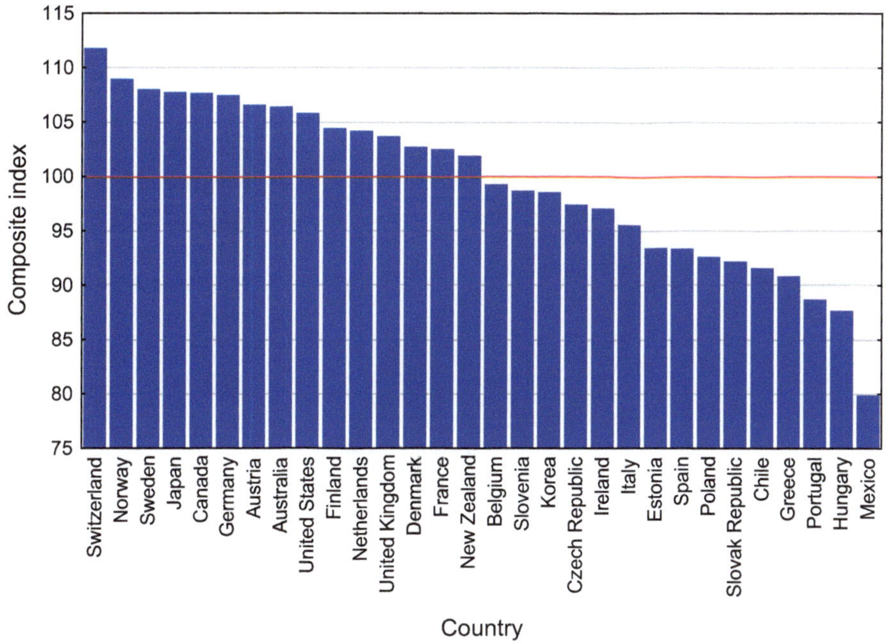

Fig. 3.11 Bar chart

Cartogram

The cartogram is a graphical representation that makes it possible to view the spatial distribution of the composite index for a series of territorial units (e.g. the Italian regions). The representation consists of a geographical map in which each area is assigned a different colour depending on the value of the composite index.

In the cartogram, the values of the composite index are divided into classes. The more detailed the representation is to be, the larger the number of classes must be. In general, classes with the same size or the same frequency (quantiles) are used. The classes, arranged in ascending order, are then associated with a series of colours with graded intensity (e.g. from light green to dark green) or with alternating intensity (e.g. red, yellow and green).

Figure 3.12 shows the cartogram of a composite index for the Italian provinces, based on the z-scores, with 7 classes of equal frequency (about 15 provinces per class).

Traveller plot

The graphical representation of traveller icons (Mazziotta and Pareto 2016a) is particularly suitable for the study of composite indices in a non-compensatory perspective, i.e. when each individual indicator is not or only partially substitutable by the others. In this case, compensation is indeed not allowed and the researcher's attention is also focused on the balance of the individual indicators.

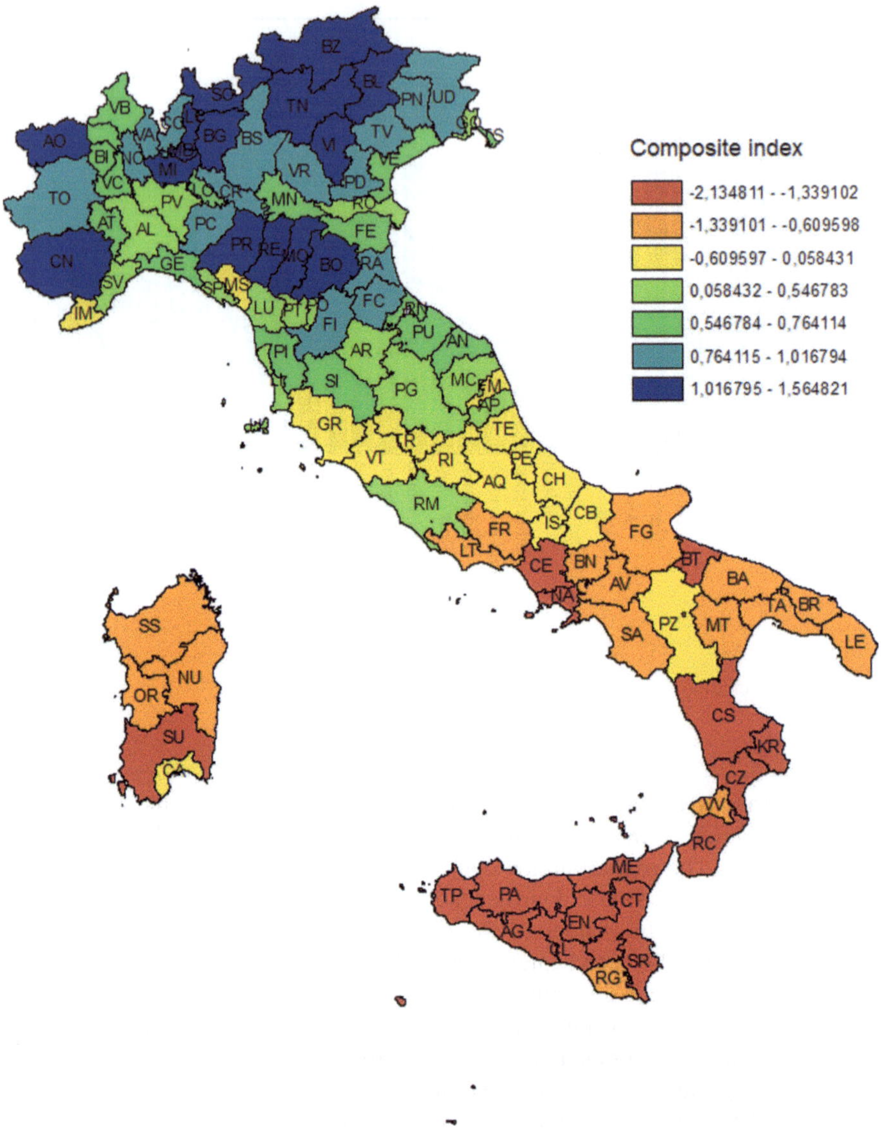

Fig. 3.12 Cartogram

The basic idea is to represent each statistical unit (e.g. a country) as a small man with a bag, where the size of the man is proportional to the mean of the normalized indicators and the size of the bag is proportional to their variability ('horizontal variability').

This representation allows us to discover interactions between 'mean' and 'variance' and to easily recognize similarities between units.

Figure 3.13 shows the traveller icons regarding the MPI for a number of countries in a given year. The height of the man represents the 'mean level' of development and the width of the bag represents the 'penalty' (i.e. imbalance in development).

Countries with a high and homogeneous level of development are represented by large men with a small bag, while countries with a low and heterogeneous level of development are represented by small men with a large bag.

Line chart

The line chart is the most suitable representation for time series and consists of the indication of time on the x-axis of a Cartesian diagram and the value of the composite index on the y-axis, so that each point represents the value of a particular unit at time t.

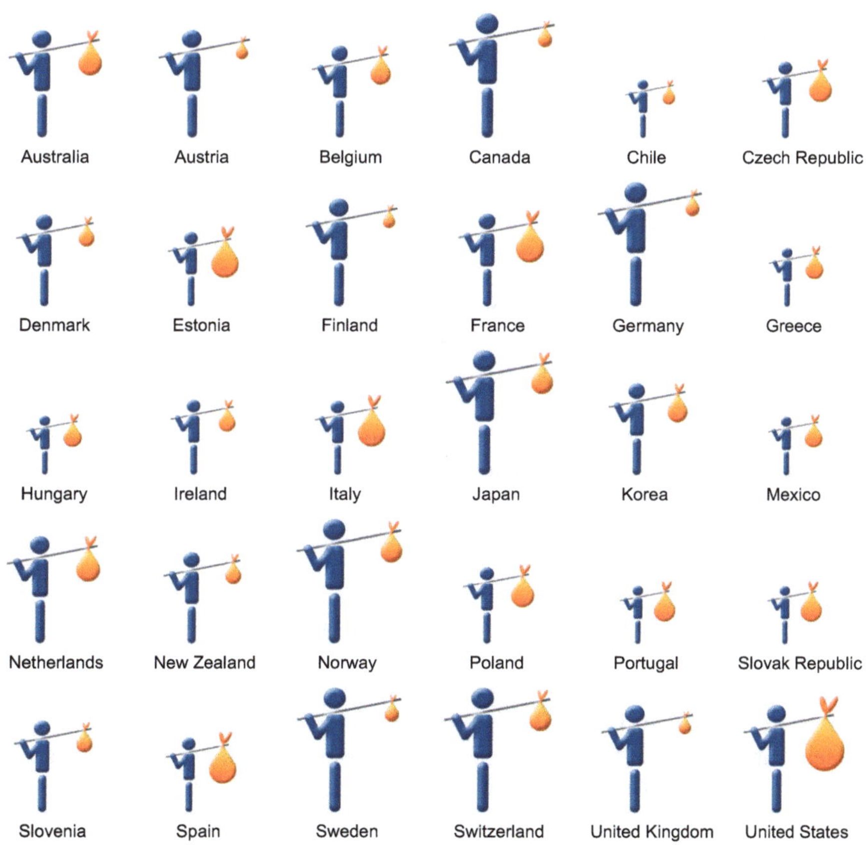

Fig. 3.13 Traveller plot

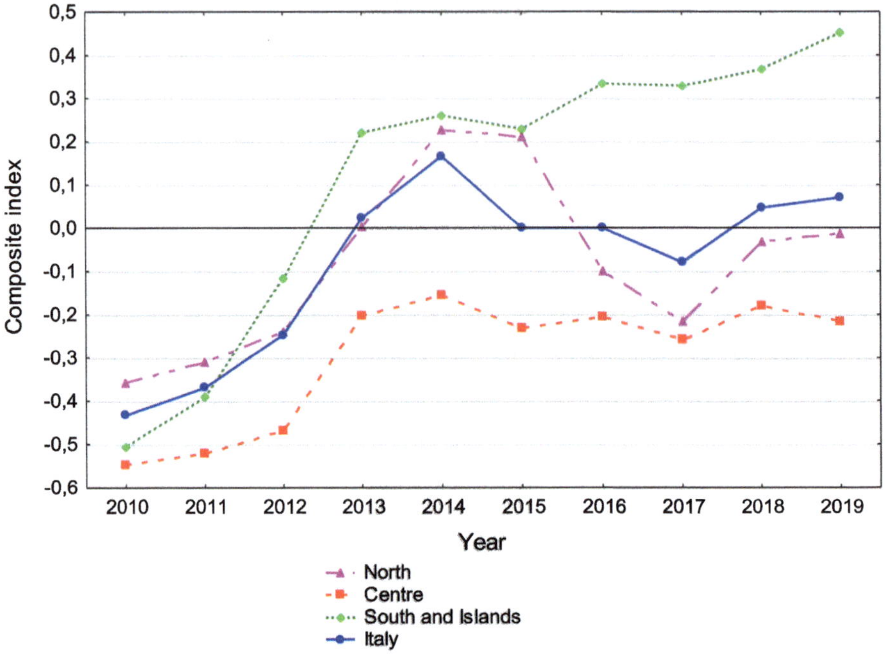

Fig. 3.14 Line chart

If we connect the points of each unit with a dashed line, we get the trend of the composite index over time. Generally, the dashed lines are drawn with different line types (solid, dashed, dotted, etc.) and/or colours to easily distinguish the different units.

Figure 3.14 shows the evolution of a composite index for the Italian geographical areas, based on a normalization with the constrained Min–Max method and a base of Italy 2015 equal to zero.

Heatmap

If the number of statistical units is high and/or the dashed lines of the line chart overlap in a confusing way, it may be more useful to visualize the time series using heatmaps.

This type of chart allows to display the values contained in a matrix with the units in rows and the time periods in columns in colours.

Heatmaps are so called because they use colour gradients to represent the relative 'temperature' of the data, with warmer colours indicating higher values and colder colours indicating lower values.

Figure 3.15a shows the evolution of a composite index for the Italian regions, with a base of 100. The most interesting aspect of this visualization is the possibility to group the units with similar trends using clustering techniques, as shown in Fig. 3.13b.

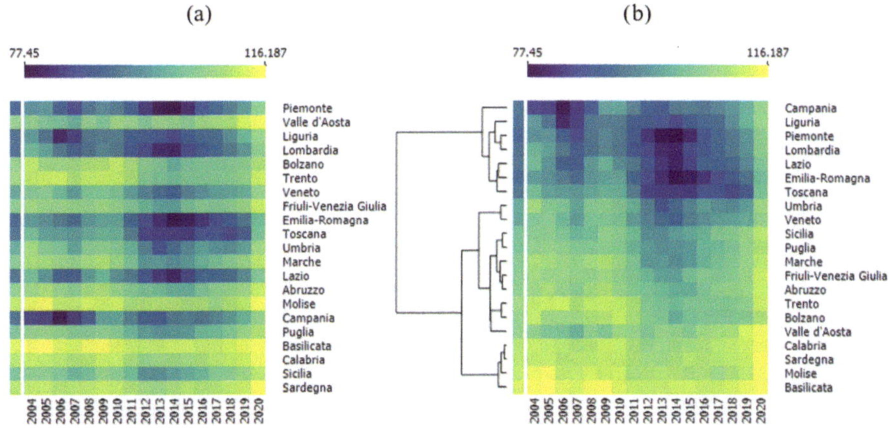

Fig. 3.15 Heatmap

Scatterplot

The scatterplot is a diagram that visualizes the relationship between two variables of interest.

It can be used for various purposes in visualizing the results. In particular:

- Visualize the relationship between the composite index and an external control variable (e.g. the relationship between a composite index of well-being and "GDP per capita").
- Visualize the relationship between the composite index and the variability of the individual components (e.g. the standard deviation of the normalized indicators). This is particularly useful in a non-compensatory or partially compensatory approach, where it is also important to assess the so-called 'horizontal variability' of the statistical units.
- Compare the composite index values in two different time periods. The further away the points are from the main diagonal (the line of equation $y = x$), the more the values differ.
- Create the so-called 'speed-acceleration' diagram, which shows the composite index calculated in the reference period (static index) on one axis and the change in the composite index in relation to a base period (dynamic index) on the other axis.

Figure 3.16 shows a speed-acceleration diagram for the Italian regions, based on the Jevons index, with the reference year 2009 and the base year 2005.

Of course, the speed-acceleration diagram can also be created for indicators other than the Jevons Index. In this case, it is sufficient to plot the composite index in the reference year on one axis and the difference between the composite index in the reference year and the composite index in the base year on the other axis.

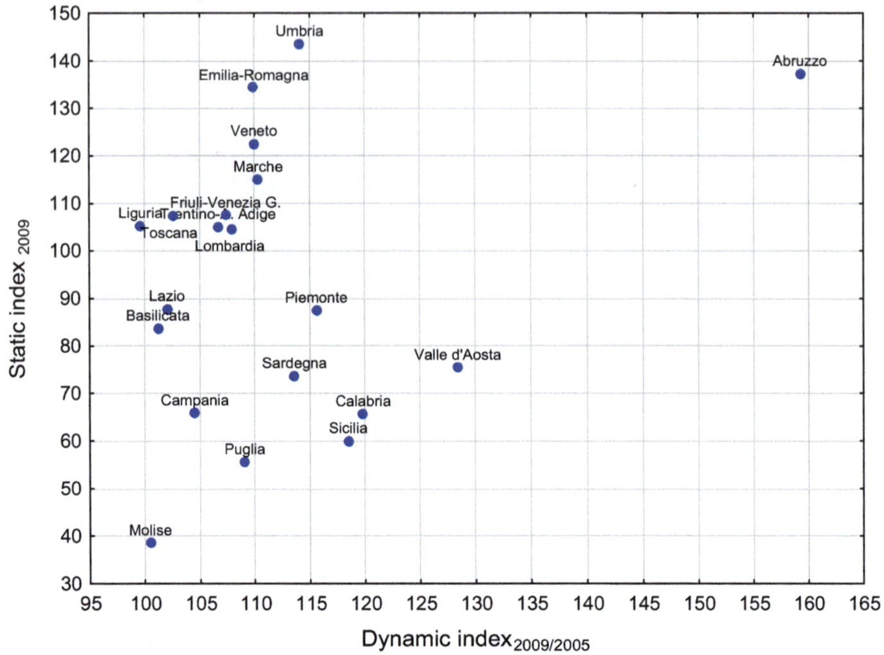

Fig. 3.16 Scatterplot

3.8 Best Practices

In general, there is no composite index that is universally valid for all areas of application, as the validity of a synthesis depends on the objectives of the research. This section proposes a basic scheme with some general guidelines to follow when constructing a composite index.

3.8.1 How to Choose the 'Best' Method

The main factors (assumptions and/or requirements) to consider when choosing the methodology for constructing a composite index are as follows:

- Type of model (reflective/formative).
- Type of indicators (substitutable/non-substitutable).[19]
- Type of aggregation (simple/complex).
- Type of comparisons (absolute/relative).

[19] Note that the term 'non-substitutable' refers to all indicators that are not fully substitutable, and thus also to those that are partially substitutable.

- Type of weights (objective/subjective).

There is not always a 'consolidated' solution and sometimes it may be necessary to sacrifice some properties to fulfil others.

Type of model. The choice of measurement model is closely linked to the selection of individual indicators. If the individual indicators are considered as the 'effect' of the latent variable, a reflective model should be used. If the individual indicators are considered the 'cause' of the latent variable, a formative model should be used. Note that the adopted model may influence the choice of aggregation method. For example, FA and PCA are the most appropriate methods for aggregating reflective indicators (interchangeable and highly correlated), but they are not appropriate for aggregating formative indicators (not interchangeable and/or poorly correlated).

Type of indicators. This is one of the main factors influencing the choice of aggregation method. If the individual indicators are considered substitutable, a compensatory approach must be chosen, otherwise it is necessary to resort to a non-compensatory approach.

Type of aggregation. The choice of aggregation method also depends on the purpose of the work and the type of audience it is aimed at (researchers or ordinary people). From this point of view, an aggregation method can be categorized as 'simple' or 'complex'. An aggregation method is considered simple if it does not take into account the relationships between individual indicators (e.g. correlations) or between statistical units (e.g. preference relationships). A simple method uses an easily understandable mathematical function (e.g. the arithmetic mean). In contrast, an aggregation method is called complex if it takes into account the relationships between individual indicators or between statistical units. A complex method uses sophisticated models or multivariate analysis techniques (e.g. PCA or NLPCA).

Type of comparisons. The choice of normalization method depends primarily on the type of comparisons to be made. All normalization methods allow spatial comparisons (comparisons between the values of different units in the same reference period), while temporal comparisons (comparisons between the values of the same unit or different units in different reference periods) may be more difficult to perform or interpret. Temporal comparisons can be considered 'absolute' or 'relative'. A composite index allows relative temporal comparisons if the values of the composite index at time t depend on one or more 'endogenous' parameters that change over time (e.g. mean and standard deviation of the individual indicators at time t). Similarly, a composite index allows absolute comparisons over time if the values of the composite index at time t depend on one or more 'exogenous' parameters that do not change over time (e.g. the minimum and maximum of each indicator for all periods considered). Ranking and standardization only allow relative comparisons, as they are based exclusively on the values of the individual indicators at time t. Other normalization methods such as Min–Max, indicization and constrained Min–Max, require that the minimum and maximum (goalposts) or the reference base are independent of time t in order to be able to make absolute comparisons (Tarantola 2008).

Type of weights. The question of choosing a weighting system for the individual indicators that reflects their different importance for the representation of the phenomenon under consideration inevitably involves the introduction of an arbitrary component. Subjective weighting can be done implicitly by assigning the same weight to all components (equal weighting) or explicitly by a group of experts who determine the weight of each individual indicator. In contrast, objective weighting can be done implicitly by choosing a normalization method that assigns a weight proportional to the variability of each indicator, or explicitly by calculating the weights using a multivariate analysis method such as PCA.

Figure 3.17 shows a flowchart for selecting the 'best' method to create a composite index with the main solutions—normalization, weighting and aggregation—for each of the choices—assumptions and requirements—made (Mazziotta and Pareto 2013).

If the phenomenon to be measured can be divided into several dimensions, each of these dimensions is represented by a subset of individual indicators. Therefore, it can be useful to create a 'partial' composite index for each dimension (or 'pillar') and then create a 'global' composite index by aggregating the partial composite indices. In this case, it is possible to take a compensatory approach within each dimension and a non-compensatory or partially compensatory approach between the different dimensions.

The most commonly used aggregation methods for substitutable indicators are linear methods, such as the arithmetic mean (simple aggregation) or PCA (complex aggregation). For non-substitutable or partially substitutable indicators, on the other hand, nonlinear methods are used, such as multiplicative functions (simple aggregation) or Multi-Criteria Analysis (MCA)[20] (complex aggregation).

If we focus on simple aggregation methods (mathematical functions), normalization depends on the type of comparisons desired and the weighting of the individual indicators.

For relative comparisons with subjective weighting (equal or different), we recommend ranking, standardization, Min–Max method with endogenous goalposts (e.g. the minimum and maximum of the distribution at time t) or constrained Min–Max method with an endogenous base (e.g. the mean of the distribution at time t). If instead we want to assign objective weights proportional to the variability of the individual indicators, we can use an indicization with an endogenous base (e.g. the mean or the maximum of the distribution at time t).

Ranking and normalization cannot be used for absolute comparisons. For subjective weighting, the Min–Max method with exogenous goalposts (e.g. the minimum and maximum of the entire time series) or the Min–Max constrained

[20] Multi-Criteria Analysis (MCA), also known in the literature as Multiple-Criteria Decision-Making (MCDM), comprises various methods, techniques and tools (with different degrees of complexity) that combine multiple objectives and criteria (or attributes) in decision problems.

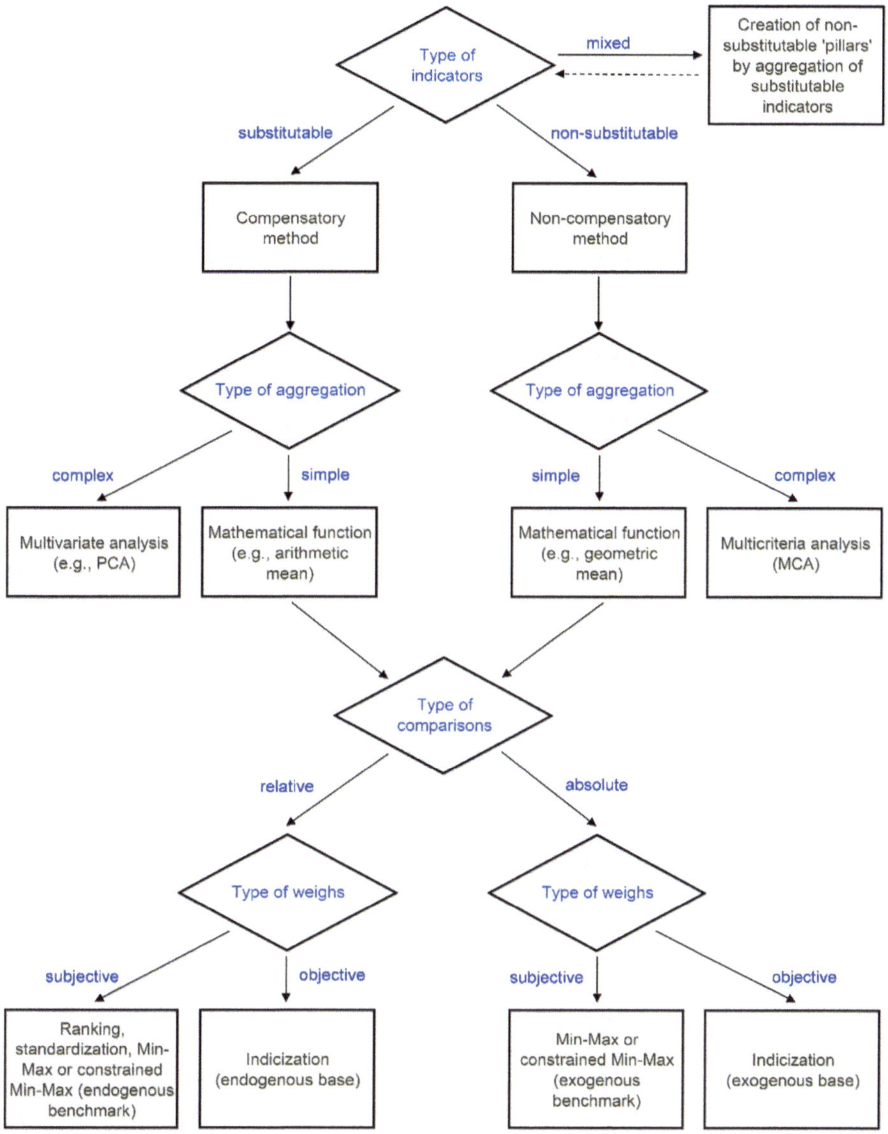

Fig. 3.17 Flowchart for choosing the 'best' method

with an exogenous base (e.g. a reference benchmark) must be used; for objective weighting, the best solution is an indicization with an exogenous base (e.g. the mean at the initial time t_0).

Of course, the correspondence grid of the chosen normalization method must reflect the desired balancing model, especially in the case of a non-compensatory or partially compensatory approach.

3.8.2 Tips and Tricks

The first criterion to be observed when constructing a composite index is the principle of parsimony. This principle states that the composite index must be as simple as possible. This enables a simple interpretation of the results, both in spatial and temporal terms. In order for the composite index to be as simple as possible, the processing of the data must be reduced to a minimum. Therefore, only a single normalization method needs to be applied to the data matrix $\mathbf{X}$[21] and no further transformation of the vector $\mathbf{Y}$ obtained from the aggregation must be performed.

Individual indicators must be expressed on an interval or ratio scale.[22] In this case, it is better not to convert them into ordinal data (e.g. by forming classes) or to use non-aggregative methods based on comparisons and rankings, such as MCA, as the 'distances' between the statistical units and thus a considerable amount of information is lost. Furthermore, in order to make comparisons between units, it is necessary to check for each indicator whether the difference between the values is significant (to establish a preference relationship) or not significant (to establish an indifference relationship), and this is very often not possible. For example, if a country has a "GDP per capita" of 32.4 thousand euros and another of 32.3 thousand euros, it is difficult to determine whether the difference of 0.1 corresponds to a preference or indifference relationship. Therefore, MCA should only be used when it is possible to determine whether there is a preference or indifference relationship between two units.

Most individual indicators should have a positive polarity to minimize polarity reversals. If the composite index is intended to measure the intensity of a multidimensional phenomenon (static index), each individual indicator should represent the 'level' of a component (e.g. "GDP per capita"). If the composite index is intended to measure the change in a multidimensional phenomenon (dynamic index), each individual indicator should represent a change (e.g. "GDP growth rate").

The number of individual indicators must not be too high in order to avoid redundancies. In the case of complex phenomena that are represented by many indicators, it can be useful to proceed step by step by grouping the homogeneous indicators and constructing partial composite indices. These indices can then be aggregated into a single global composite index.

The choice of measurement model depends on the type of phenomenon under investigation. If the statistical unit is a geographical area and the individual indicators concern aspects such as health, income, employment, services and the environment, which are considered to be the cause of the phenomenon to be measured (e.g. well-being), a formative model is generally chosen. However, if the individual indicators are seen as the effect of an underlying phenomenon that cannot be measured directly (e.g. corruption), a reflective model is used. Similarly, if

[21] All individual indicators must be normalized in the same way.

[22] Note that any criterion for 'quantifying' variables measured on an ordinal scale is arbitrary (Casacci and Pareto, 2015).

the statistical unit is an individual and the individual indicators relate to aspects such as education, income and occupational prestige that are considered to be the cause of the phenomenon being measured (e.g. SSE), a formative model is chosen. However, if the individual indicators are the answers to a questionnaire on an underlying phenomenon that cannot be measured directly (e.g. intelligence), a reflective model is used.

In a reflective model, individual indicators can be selected on the basis of the correlation matrix and they should be highly correlated. In a formative model, the selection of indicators should not be based on the correlation matrix, as this could lead to the exclusion of valid measures.

When choosing the normalization method, the type of weighting (implicit and/or explicit) to be applied and the type of comparisons required must be taken into account. In the case of a single data matrix **X**, a 'static' analysis with relative comparisons is sufficient, whereas in the case of a sequence of matrices $\mathbf{X}^1$, ..., $\mathbf{X}^p$, which refer to p periods, a 'dynamic' analysis with absolute comparisons is appropriate. Of course, in order to make correct comparisons, all matrices must consist of the same set of individual indicators and the reference values for normalizing the data (e.g. the goalposts or the base of the index numbers) must remain constant over time.

In this context, it should be recalled that PCA does not allow comparisons over time, as the variance explained and the weights calculated with PCA would be different for each matrix $\mathbf{X}^t$ ($t = 1, ..., p$) and the results obtained would therefore not be easily comparable. This can be a major problem if the composite index needs to be calculated and evaluated over time. The problem could be solved using Dynamic Factor Analysis methods, such as the STATIS method—*Structuration des tableaux à trois indices de la statistique* (Lavit et al. 1994), but the composite index would have to be recalculated for the entire series each time a new time period is available. It should also be noted that PCA cannot be applied to matrices containing the values of a set of indicators (columns) for different time periods (rows), as the correlations between individual indicators must be calculated on the basis of independent observations (e.g. persons or geographical areas).

In order to weight individual indicators equally or explicitly assign subjective weights to them, it is necessary to remove the 'variability effect' as much as possible to avoid 'implicit' weighting, which could distort the results.

In the case of a non-compensatory or partially compensatory approach, it is particularly important to define what is meant by balance and to choose a normalization method that can reproduce the desired model.

The aggregation method must be consistent with the chosen measurement model (reflective/formative). The synthesis must be compensatory, partially compensatory or non-compensatory, depending on the degree of compensability or substitutability of the individual indicators.

Finally, if different methods are considered valid, it may be useful to perform an influence and/or uncertainty analysis to select the most robust method.

3.9 Conclusion

Some researchers believe that it is useless and misleading to construct composite indicators because they claim that composite indicators can be constructed in a thousand different ways, with a thousand different results.

This claim hides a rather superficial knowledge of 'statistics' for composite indicators, because out of a thousand different composite indices, only a handful will be correct (because they fulfil all the requirements and assumptions of the theoretical framework) and give similar results.

The methods of normalization and aggregation of data are not interchangeable, as some believe, and each has very specific characteristics.

Therefore, it is important to justify the choices made and to be able to refute any criticism.

Anyone who presents a composite index to the public is often told: "But why didn't you use the index numbers instead of the *z*-scores?" or "But why didn't you use the arithmetic mean instead of the geometric mean?" or, even worse, "But why didn't you use PCA?".

Anyone who has constructed a composite index must be able to answer these questions and explain, for example, that "The index numbers were not used because we wanted to remove implicit weighting", that "The arithmetic mean was not used because a non-compensatory approach was chosen" or that "PCA was not used because the measurement model was formative and not reflective".

The perfect composite index does not exist, but only with a rigorous statistical-methodological approach, such as the one presented in this chapter, will it be possible to construct the best composite index, since all possible alternatives were worse.

References

Annoni P, Bolsi P (2020) The regional dimension of social progress in Europe: presenting the new EU social progress index. WP 06/2020. Publications Office of the European Union, Luxembourg

Bargiela A, Hartley JK (1993) Orthogonal linear regression algorithm based on augmented matrix formulation. Comput Oper Res 20:829–836

Booysen F (2002) An overview and evaluation of composite indices of development. Soc Indic Res 59:115–151

Box GEP, Cox DR (1964) An analysis of transformations. J R Stat Soc B-26:211–243

Bollen KA, Lennox R (1991) Conventional wisdom in measurement: a structural equation perspective. Psychol Bull 110:305–314

Casacci S (2023) Categorical principal component analysis. In: Maggino F (ed) Encyclopedia of quality of life and well-being research. Springer, Cham, pp 677–678

Casacci S, Pareto A (2015) Methods for quantifying ordinal variables: a comparative study. Qual Quant 49:1859–1872

Casacci S, Pareto A (2021) A nonlinear multivariate method for constructing profiles of country performance. Int J Multicriteria Decis Mak 8:295–312

Casadio Tarabusi E, Guarini G (2013) An unbalance adjustment method for development indicators. Soc Indic Res 112:19–45

Cattell RB (1966) The scree test for the number of factors. Multiv Behav Res 1:245–276

De Muro P, Mazziotta M, Pareto A (2011) Composite indices of development and poverty: an application to MDGs. Soc Indic Res 104:1–18

Diamantopoulos A, Riefler P, Roth KP (2008) Advancing formative measurement models. J Bus Res 61:1203–1218

Diamantopoulos A, Winklhofer HM (2001) Index construction with formative indicators: an alternative to scale development. J Mark Res 38:269–277

Draper NR, Smith H (1998) Applied regression analysis, 3rd edn. Wiley, New York

Dunteman GH (1989) Principal components analysis. Sage Publications, Newbury Park

Freudenberg M (2003) Composite indicators of country performance: a critical assessment. OECD-DSTI/IND, Paris

Gifi A (1990) Nonlinear multivariate analysis. Wiley, New York

Giudici P, Avrini F (2002) Modelli statistici per la costruzione di indicatori della qualità della vita: aspetti metodologici. Rivista di Statistica Ufficiale 1:61–80

Harbison FH, Maruhnic J, Resnick JR (1970) Quantitative analyses of modernization and development. Princeton University Press, Princeton, NJ

Jacobs R, Smith P, Goddard M (2004) Measuring performance: an examination of composite performance indicators. The University of York—CHE Technical Paper Series, 29, York

Kaiser HF (1960) The application of electronic computers to factor analysis. Educ Psychol Meas 20:141–151

Lavit C, Escoufier Y, Sabatier R, Traissac P (1994) The ACT (STATIS method). Comput Stat Data Anal 18:97–119

Linting M, Meulman JJ, Groenen PJF, Van der Kooij AJ (2007) Nonlinear principal components analysis: Introduction and application. Psychol Methods 12:336–358

Luzzati T, Gucciardi G (2015) A non-simplistic approach to composite indicators and rankings: an illustration by comparing the sustainability of the EU Countries. Ecol Econ 113:25–38

Luzzati T, Gucciardi G (a cura di) (2013) La sostenibilità delle regioni italiane: la classifica di IRTA Leonardo. IRTA Leonardo, Pisa

MacKenzie SB, Podsakoff PM, Podsakoff NP (2011) Construct measurement and validation procedures. MIS Q 35:293–334

Maggino F, Zumbo BD (2012) Measuring the quality of life and the construction of social indicators. In: Land KC et al (eds) Handbook of social indicators and quality of life research. Springer, Dordrecht, pp 201–238

Mazziotta M, Pareto A (2012) A non-compensatory approach for the measurement of the quality of life. In: Maggino F, Nuvolati G (eds) Quality of life in Italy: research and reflections. Springer, New York, pp 27–40

Mazziotta M, Pareto A (2013) Methods for constructing composite indices: one for Allor all for one? Ital J Econ Demogr Stat Stud LXVII 2:67–80

Mazziotta M, Pareto A (2015) Comparing two non-compensatory composite indices to measure changes over time: a case study. Statistika 95:44–53

Mazziotta M, Pareto A (2016a) A well-being index based on the weighted product method. In: Alleva G, Giommi A (eds) Topics in theoretical and applied statistics, studies in theoretical and applied statistics. Springer, Cham, pp 253–259

Mazziotta M, Pareto A (2016b) On a generalized non-compensatory composite index for measuring socio-economic phenomena. Soc Indic Res 127:983–1003

Mazziotta M, Pareto A (2017) Synthesis of indicators: the composite indicators approach. In: Maggino F (ed) Complexity in society: from indicators construction to their synthesis. Springer, Cham, pp 159–191

Mazziotta M, Pareto A (2019) Use and misuse of PCA for measuring well-being. Soc Indic Res 142:451–476

Mazziotta M, Pareto A (2021) Everything you always wanted to know about normalization (but were afraid to ask). Ital J Econ Demogr Stat Stud LXXV(1):41–52

Mazziotta M, Pareto A (2022a) Composite indices construction: the performance interval approach. Soc Indic Res 161:541–551

Mazziotta M, Pareto A (2022) Normalization methods for spatio-temporal analysis of environmental performance: revisiting the Min–Max method. Environmetrics 33:5
Mazziotta M, Pareto A (2024) Principal component analysis for constructing socio-economic composite indicators: theoretical and empirical considerations. SN Soc Sci 4:114. https://doi.org/10.1007/s43545-024-00920-x
McGranahan D, Pizarro E, Richard C (1985) Measurement and analysis of socio-economic development. United Nations Research Institute for Social Development, Geneva
Mishra SK (2007) A comparative study of various inclusive indices and the index constructed by the principal components analysis. North-Eastern Hill University- MPRA Paper 3377, Shillong
OECD (2008) Handbook on constructing composite indicators. Methodology and user guide. OECD Publications, Paris
Pareto A (2023) A robust method for regression and correlation analysis of socio-economic indicators. Qual Quant 57:5035–5053
Pareto A (2024) On a correlation coefficient based on the L_1-norm. Statistical papers.https://doi.org/10.1007/s00362-024-01616-3
Rousseeuw PJ, Leroy AM (1987) Robust Regression and Outlier Detection. Wiley, New York
Saisana M, Tarantola S, Saltelli A (2005) Uncertainty and sensitivity techniques as tools for the analysis and validation of composite indicators. J R Stat Soc A 168:307–323
Saltelli A (2007) Composite Indicators between analysis and advocacy. Soc Indic Res 81:65–77
Salzman J (2003) Methodological choices encountered in the construction of composite indices of economic and social well-being. Center for the Study of Living Standards, Ottawa
Silvio-Pomenta JF (1973) Typological study using the Wroclaw taxonomic method (a study of regional disparities in Venezuela). UNESCO-SHC/WS/316, Paris
Sobol IM (1993) Sensitivity analysis for non-linear mathematical models. Math Model Comput Exp 1:407–414
Spencer NH (2014) Essentials of multivariate data analysis. CRC Press, Boca Raton, FL
Tarantola S (2008) European Innovation Scoreboard: strategies to measure country progress over time. JRC-EUR 23526 EN, Luxembourg
Terzi S, Moroni L (2004) Graduatorie della qualità della vita e loro sensibilità al pre-trattamento delle variabili che la definiscono: alcune critiche al dossier de Il Sole 24 Ore. In: Quaderni di Statistica, vol 6. Roma Tre University, Rome
Terzi S, Otoiu A, Grimaccia E, Mazziotta M, Pareto A (2021) Open issues in composite indicators. A starting point and a reference on some state-of-the-art issues. Roma TrE-Press, Rome
USAID (2014) Design and Use of Composite Indices in Assessments of Climate Change Vulnerability and Resilience. U.S. Agency for International Development, Washington

Applications to Real Data 4

4.1 Introduction

This chapter presents some applications to real data for the construction of composite indices by different normalization and aggregation methods and the comparison of the results obtained.

Section 4.2 is dedicated to the 'static' analysis, i.e. the spatial analysis of the phenomenon under study in a given time period, and is based on a data matrix of the type $\mathbf{X}_{n,m} = \{x_{ij}\}$, with n rows (statistical units) and m columns (individual indicators). The aim is therefore to make relative comparisons between the different units.

Section 4.3 introduces time and deals with the 'dynamic' analysis, i.e. the spatio-temporal analysis based on a sequence of matrices of type $\mathbf{X}^t_{m,n}$ ($t = 1, \ldots, p$), where t is the observation period. In this case, the interest is mainly focused on the methods that allow absolute comparisons.

4.2 Spatial Analysis

The first application presented in this section concerns the measurement of well-being in the OECD countries of the European Union based on a set of individual indicators of the *Better Life Index* (OECD 2020). To this end, some compensatory and non-compensatory indices are constructed and compared, and their degree of robustness is assessed through an influence analysis. Finally, an alternative approach based on performance intervals is shown.

The second application deals with the aggregation of a set of indicators from the third Italian report on "Equitable and Sustainable Well-being (BES)" (Istat 2015) and the study of their relationship with GDP per capita.

The third application is dedicated to the construction of development profiles for the countries of the world based on some socioeconomic indicators from the

M. Mazziotta and A. Pareto, *Statistics for Composite Indicators*, SxI - Springer for Innovation / SxI - Springer per l'Innovazione 18,
https://doi.org/10.1007/978-3-031-97978-1_4

"Human Development Report 2020" (UNDP 2020). In this case, the non-linear approach based on the multivariate normalization of the individual indicators is compared with the classical approach based on the univariate normalization by transformation into z-scores.

4.2.1 Measuring Well-Being

Let us consider the dataset in Table 4.1, which contains five individual indicators of well-being for the OECD countries of the European Union in 2023. The dimensions analysed are: health, education, work, wealth and environment. The selected indicators are as follows:

- X1. Life expectancy (years);
- X2. Educational attainment (%);
- X3. Employment rate (%);
- X4. Disposable income of households ($);
- X5. Air pollution (micrograms PM2.5 per m^3).

The source is the *Better Life Index* (BLI) and the data can be found at: https://www.oecdbetterlifeindex.org.

Table 4.1 also contains the most important summary statistics of the individual indicators for the preliminary exploratory analysis. The indicators have very different magnitudes, as they are expressed in different units of measurement, and all have a positive polarity (e.g. the higher the life expectancy, the higher the well-being), with the exception of X5, which has a negative polarity (the higher the air pollution, the lower the well-being).

As far as variability is concerned, the standard deviation (Std) is not directly comparable and it is therefore necessary to resort to the study of the coefficient of variation (CV). The CV varies from 3.1% for life expectancy (X1)—with relatively close minimum and maximum (75.5 vs. 83.9)—to 35.5% for air pollution (X5)—with a maximum corresponding to four times the minimum (22.8 vs. 5.5). This means that in the absence of a normalization method that adjusts for the variability effect, X1 will have the lowest implicit weight and X5 the highest implicit weight.

X1, X2 and X3 have a left-skewed distribution (negative skewness) where the highest frequencies are concentrated at the end of the range. Indeed, these indicators have a median that is significantly higher than the mean. In contrast, X4 and X5 have a right-skewed distribution (positive skewness), with a higher frequency of the lowest values. In this case, both indicators have a median that is lower than the mean. This result is confirmed by the skewness index, which is negative for X1, X2 and X3 and positive for X4 and X5.

The frequency histograms of the individual indicators are shown in the matrix diagram in Fig. 4.1. As we can see, no indicator is normally distributed and some distributions are multimodal. The diagram also shows the scatterplots for each pair of indicators.

Table 4.1 Individual indicators of well-being (Year 2023)

Country/statistic	X1	X2	X3	X4	X5
Austria	82.0	86	72	37,001	12.2
Belgium	82.1	80	65	34,884	12.8
Czechia	79.3	94	74	26,664	17.0
Denmark	81.5	82	74	33,774	10.0
Estonia	78.8	91	74	23,784	5.9
Finland	82.1	91	72	33,471	5.5
France	82.9	81	65	34,375	11.4
Germany	81.4	86	77	38,971	12.0
Greece	81.7	76	56	20,791	14.5
Hungary	76.4	86	70	21,026	16.7
Ireland	82.8	85	68	29,488	7.8
Italy	83.6	63	58	29,431	15.9
Latvia	75.5	89	72	19,783	12.7
Lithuania	76.4	94	72	26,976	10.5
Luxembourg	82.7	74	67	44,773	10.0
Netherlands	82.2	81	78	34,984	12.2
Poland	78.0	93	69	23,675	22.8
Portugal	81.8	55	69	24,877	8.3
Slovak Republic	77.8	92	68	21,149	18.5
Slovenia	81.6	90	71	25,250	17.0
Spain	83.9	63	62	27,155	10.0
Sweden	83.2	84	75	33,730	5.8
Min	75.5	55.0	56.0	19,783.0	5.5
Max	83.9	94.0	78.0	44,773.0	22.8
Mean	80.8	82.5	69.5	29,364.2	12.3
Median	81.8	85.5	70.5	28,293.0	12.1
Std	2.5	10.4	5.5	6576.0	4.3
CV %	3.1	12.6	7.9	22.4	35.5
Skewness	– 0.84	– 1.19	– 0.81	0.40	0.42
Kurtosis	– 0.65	0.62	0.18	– 0.62	– 0.25

The bivariate distributions clearly shows the presence of some outliers and the existence of particularly weak relationships—increasing and decreasing, linear and non-linear—between the various indicators. For example, life expectancy (X1) and educational attainment (X2) show a decreasing, curvilinear relationship, which corresponds to a parabola with downward concavity, while educational attainment (X2) and employment rate (X3) show an increasing but rather weak linear relationship.

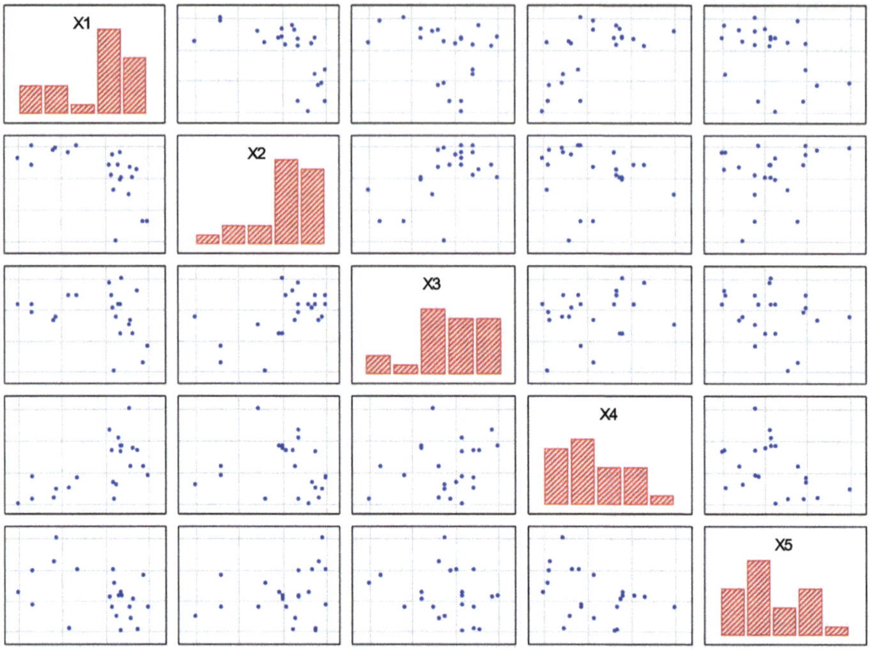

Fig. 4.1 Matrix plot of the individual indicators of well-being (Year 2023)

Table 4.2 shows the correlation and rank correlation matrices of the individual indicators. The most strongly correlated indicators are life expectancy (X1) and disposable income of households (X4) with $r = 0.615$; while the lowest correlation is between educational attainment (X2) and disposable income of households (X4) with $r = -0.136$.

The linear correlation between life expectancy (X1) and educational attainment (X2) is -0.587, while the rank correlation rises to -0.701 due to the strong decreasing non-linear relationship between the two indicators.

Overall, the indicators are not strongly correlated, which confirms that they cannot be considered as an effect of a latent factor called 'well-being', as in the reflective model, but rather as a cause. Therefore, the composite index must be constructed according to a formative approach.[1]

The lack of strong linear relationships between the individual indicators is confirmed by PCA, the results of which are presented in Table 4.3, which shows for each principal component the percentage of variance explained (absolute and cumulative) and the weights of the individual indicators.

[1] However, it should be noted that the choice of model does not depend on the observed correlations, but on the theoretical framework. Therefore, a formative approach should be chosen even for highly correlated indicators (see e.g. the HDI).

Table 4.2 Relationships between the individual indicators of well-being (Year 2023)

Individual indicator	X1	X2	X3	X4	X5
Bravais-Pearson correlation coefficient					
X1	1.000				
X2	– 0.587	1.000			
X3	– 0.290	0.541	1.000		
X4	0.615	– 0.136	0.223	1.000	
X5	– 0.389	0.213	– 0.228	– 0.393	1.000
Spearman rank correlation coefficient					
X1	1.000				
X2	– 0.701	1.000			
X3	– 0.339	0.484	1.000		
X4	0.582	– 0.320	0.203	1.000	
X5	– 0.420	0.264	– 0.229	– 0.384	1.000

Table 4.3 PCA of the individual indicators of well-being (Year 2023)

Principal component	Variance explained (%)		Weight				
	Value	Cumulative	X1	X2	X3	X4	X5
1	44.92	44.92	– 0.615	0.493	0.210	– 0.439	0.376
2	31.78	76.70	– 0.009	0.396	0.694	0.437	– 0.412
3	13.28	89.98	0.247	0.245	-0.041	0.557	0.753
4	6.22	96.20	– 0.154	– 0.710	0.605	– 0.014	0.325
5	3.80	100.00	0.733	0.187	0.325	– 0.554	0.126

The table shows that the first principal component explains only 44.92% of the total variance. In order to reproduce at least 90% of the original information, the first three components must be taken into account. Another important aspect concerns the signs of the weights of the individual indicators in the first principal component, which do not match their polarities. This is due to the fact that PCA is based exclusively on the matrix of observed correlations and 'ignores' the polarities defined by the user.

As we know, in a correct composite index, the individual indicators with positive polarity must be positively weighted, while the indicators with negative polarity must be negatively weighted.[2]

So if we want to use the first principal component as a composite index, the first four indicators (X1–X4) should be positively weighted and the last one (X5)

[2] If all indicators are normalized to a positive polarity—whereby the polarity of the 'negative' indicators is reversed—each of them must have a positive weight.

negatively weighted. This would only be possible if X1–X4 were positively correlated with each other and negatively with X5. In this case, life expectancy (X1) and educational attainment (X2) are negatively correlated ($r = -0.587$), although they both have a positive polarity. Consequently, their signs on the first principal component are reversed: life expectancy (X1) has a negative weight (− 0.615) and educational attainment (X2) a positive weight (0.493). The same applies to other pairs of variables, since condition (3.10) in Sect. 3.5.3 is not always fulfilled. Therefore, the first principal component cannot be used as a composite index, regardless of the amount of variance explained.

Assuming that the five dimensions of well-being—and thus the five individual indicators that represent them—have the same importance, the composite index must be constructed in such a way that the indicators to be aggregated are assigned the same weight (implicit and explicit).

Table 4.4 shows the scores and rankings of the countries according to five composite indices of well-being:

- MZ. Arithmetic mean of standardized indicators (z-score);
- M01. Arithmetic mean of indicators re-scaled to [0, 1] (Min–Max);
- MPI. Mazziotta-Pareto Index with negative penalty;
- JS. Static Jevons Index (mean = 100);
- WR. Wroclaw Index.

MZ and M01 are compensatory composite indices, while MPI, JS and WR are partially compensatory composite indices.

For MZ, the reference value is the mean value, which is equated with zero. Countries with a positive score have overall standardized values of the individual indicators that are above the mean, while countries with a negative score have overall values that are below the mean. In addition, the deficit of one standard deviation in one individual indicator is compensated for by the surplus of one standard deviation in another indicator.

In M01, the reference values are the minimum and maximum, which are set to 0 and 1 respectively. However, the interpretation of the values is much more difficult because, as noted in Sect. 3.4.2, there is no common mean (each normalized indicator has a different mean) and, as shown in Sect. 3.4.4, the correspondences established between the different indicators can also be very distorted. For example, in this case, the normalized educational attainment (X2) has a mean value of 0.71, while the normalized disposable income of households (X4) has a mean value of 0.38. Therefore, a normalized score of 0.5 for X2 corresponds to a low original value, while the same score for X4 corresponds to a high original value.

As for the partially compensatory indices, both MPI and JS are expressed on a scale whose mean value is set to 100. However, the normalized values of the individual indicators in MPI are expressed as a standard deviation, while in JS index they are expressed as a percentage. Moreover, in MPI, the horizontal variability of the individual indicators is explicitly penalized by the CV. In contrast, in JS there is an implicit penalization of the geometric mean (see Table 3.5). Finally,

Table 4.4 Composite indices of well-being (Year 2023)

Country	MZ		M01		MPI		JS		WR	
	Value	Rank	Value	Rank	Value	Rank	Value	Rank	Value	Rank
Austria	0.49	5	0.72	5	104.8	5	106.8	8	0.37	4
Belgium	0.04	11	0.60	11	100.0	11	100.9	12	0.53	10
Czechia	– 0.04	13	0.58	12	98.9	12	95.1	15	0.64	13
Denmark	0.45	7	0.70	7	104.4	6	108.5	7	0.38	6
Estonia	0.29	9	0.65	9	102.0	9	114.0	3	0.58	11
Finland	0.79	2	0.80	2	107.8	2	124.1	1	0.32	3
France	0.17	10	0.64	10	101.3	10	103.5	11	0.51	9
Germany	0.69	3	0.77	3	106.6	3	109.5	6	0.32	2
Greece	– 0.91	22	0.36	22	90.0	22	85.2	22	0.91	22
Hungary	– 0.73	20	0.39	21	92.0	19	87.8	18	0.83	19
Ireland	0.36	8	0.69	8	103.4	8	110.3	5	0.47	8
Italy	– 0.73	21	0.41	19	91.2	21	87.4	19	0.85	20
Latvia	– 0.52	17	0.44	17	93.5	17	92.5	17	0.81	17
Lithuania	– 0.04	12	0.57	14	98.7	14	103.6	10	0.65	14
Luxembourg	0.47	6	0.72	6	103.6	7	110.6	4	0.44	7
Netherlands	0.57	4	0.74	4	105.3	4	106.0	9	0.38	5
Poland	– 0.70	19	0.40	20	91.6	20	85.9	21	0.87	21
Portugal	– 0.42	16	0.48	16	94.2	16	96.5	13	0.77	16
Slovak Republic	– 0.65	18	0.41	18	92.7	18	87.1	20	0.83	18
Slovenia	– 0.08	14	0.57	13	98.7	13	93.0	16	0.64	12
Spain	– 0.36	15	0.50	15	95.0	15	95.7	14	0.75	15
Sweden	0.85	1	0.81	1	108.3	1	122.4	2	0.30	1

WR can be read as the 'normalized' distance of countries from the ideal unit (the ideal country with the best values of the five indicators); therefore, the ranking of countries by well-being is obtained by ordering the units from the lowest to the highest value of WR.

Figure 4.2 shows the matrix plot of the five composite indices, which contains the frequency histograms of the univariate distributions and the scatterplots of the bivariate distributions.

As we can see, the compensatory composite indices (MZ and M01) provide similar results, while the differences between the partially compensatory indices (MPI, JS and WR) are more pronounced, due to the different types of penalization. The index that differs the most from the others is JS, as it is based on a normalization of the individual indicators that does not free them from the 'variability' effect. The differences between the rankings according to the different composite indices are quantified in Table 4.5, which shows for each pair of composite indices

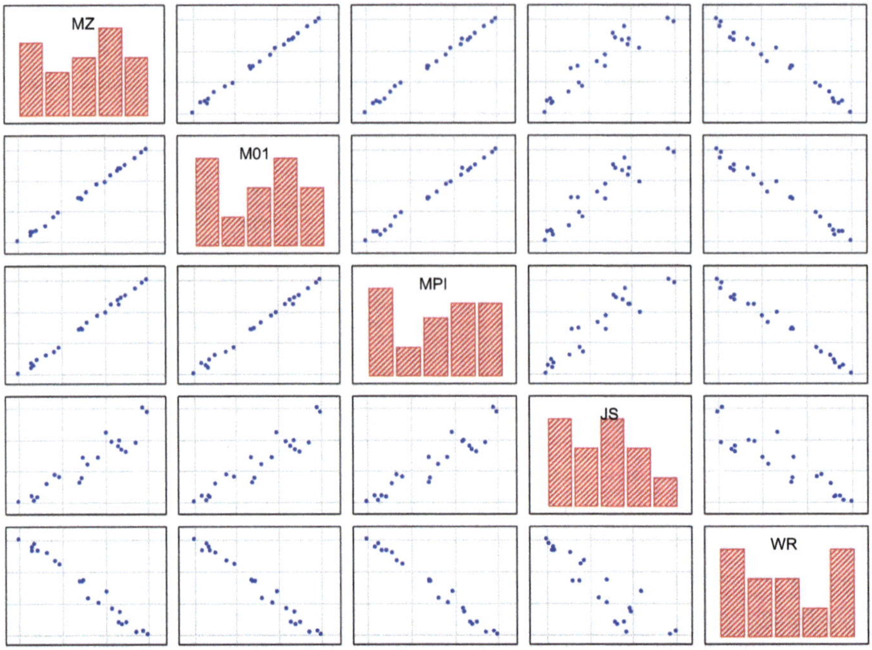

Fig. 4.2 Matrix plot of the composite indices of well-being (Year 2023)

the rank correlation and the mean absolute difference of rank (MAD). The latter is also referred to as mean of the absolute differences of rank and is defined as follows:

$$\mathrm{MAD}_{I_j,I_k} = \frac{\sum_{i=1}^{n} \left|\mathrm{rank}(I_{ij}) - \mathrm{rank}(I_{ik})\right|}{n}$$

where $\mathrm{rank}(I_{ij})$ and $\mathrm{rank}(I_{ik})$ are the positions of unit i according to the composite index j and k, respectively. The analysis of the mean absolute difference of rank shows by how many positions the countries shift on average in the rankings when we switch from one method to another.

The Spearman index between MZ and M01 is 0.993, with MAD = 0.36. This means that, on average, countries only shift by 0.36 positions when switching from one ranking to the other, so that most of them retain their position.

For the partially compensatory indices, the MPI and WR results are the most similar, with a rank correlation of − 0.992 and a MAD of 0.55 positions.

In contrast, the JS ranking shows an average gap of more than two positions compared to the other rankings. In particular, the largest divergence is between JS and WR, with $\rho = -0.889$ and MAD = 2.36.

Table 4.6 shows the results of the Influence Analysis (see Sect. 3.6.1) for the five well-being composite indices. The analysis makes it possible to assess, for

Table 4.5 Relationships between composite indices of well-being (Year 2023)

Composite index	MZ	M01	MPI	JS	WR
Spearman rank correlation coefficient					
MZ	1.000				
M01	0.993	1.000			
MPI	0.994	0.994	1.000		
JS	0.924	0.913	0.912	1.000	
WR	– 0.985	– 0.989	– 0.992	– 0.889	1.000
Mean absolute difference of rank					
MZ	0.00				
M01	0.36	0.00			
MPI	0.36	0.27	0.00		
JS	2.00	2.09	2.18	0.00	
WR	0.82	0.73	0.55	2.36	0.00

each of them, the average shift of countries in the ranking resulting from the elimination of an individual indicator from the initial group.

Starting from the five individual indicators ($m = 5$), $L = 5$ replications were carried out, removing a different indicator each time and calculating the values of

Table 4.6 Mean and standard deviation of the absolute differences of rank of countries, corresponding to the elimination of an individual indicator (Year 2023)

Individual indicator	MZ	M01	MPI	JS	WR
Mean of the absolute differences of rank					
X1	1.82	2.00	1.91	0.27	1.36
X2	1.82	1.91	1.45	1.09	1.27
X3	2.09	2.09	2.36	0.64	1.64
X4	1.91	1.64	1.55	1.45	1.27
X5	2.00	1.64	1.64	3.82	1.36
Mean	1.93	1.85	1.78	1.45	1.38
Std	0.11	0.19	0.33	1.25	0.13
Standard deviation of the absolute differences of rank					
X1	1.50	1.57	1.59	0.45	1.26
X2	1.43	1.44	1.53	0.90	1.29
X3	1.35	1.41	1.43	0.83	1.33
X4	1.90	1.82	1.64	1.64	1.09
X5	1.51	1.52	1.52	2.74	1.40
Mean	1.54	1.55	1.54	1.31	1.28
Std	0.19	0.15	0.07	0.81	0.10

the composite indices on the basis of the four remaining indicators ($L - 1 = 4$). For each replication l, the rankings of the countries were compiled according to the different methods and the absolute differences of rank between the position in the original ranking and the position in the ranking according to the $L - 1$ indicators were calculated for each country. Finally, the mean and standard deviation of the absolute differences of rank were calculated.

The table shows that the most robust index is MZ. In this case, the countries shift on average by about two positions (1.82 if X1 or X2 is eliminated and 2.09 if X3 is eliminated), resulting in a mean of 1.93 and a standard deviation of 0.11. The variabilities of the shifts are also quite stable (1.35 if X3 is eliminated and 1.90 if X4 is eliminated), with a mean of 1.54 and a standard deviation of 0.19.

In contrast, the least robust index is JS, with a standard deviation of the mean absolute differences of rank of 1.25. In this case, the ranking remains almost unchanged when X1 is eliminated (MAD = 0.27), while the countries shift on average by almost four positions (MAD = 3.82) when X5 is eliminated. This result confirms that the use of indicization, where the variability effect is not adjusted for, assigns the lowest implicit weight to X1 and the highest implicit weight to X5.

The influence of the various individual indicators on the composite indices can be visualized by the bar chart in Fig. 4.3.

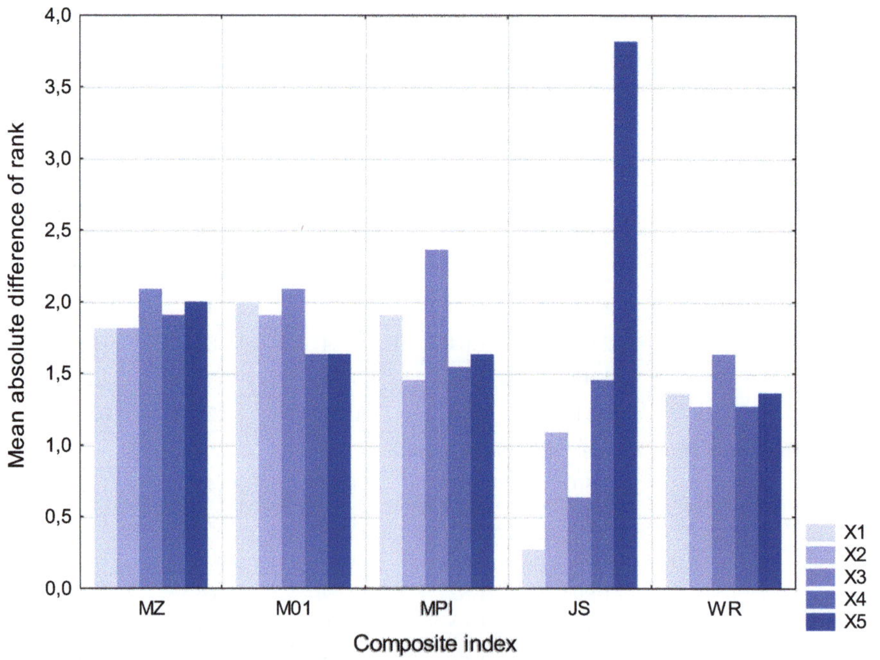

Fig. 4.3 Mean of the absolute differences of rank of countries, corresponding to the elimination of an individual indicator (Year 2023)

If a distribution is uniform (zero variability), the influence of the individual indicators (and thus their implicit weight) is the same. The more jagged the distribution is, the more different the influence of the individual indicators is (some have a large implicit weight and others a small implicit weight). Therefore, among the compensatory composite indices, MZ is more robust than M01, while among the partially compensatory composite indices, WR is the most robust and JS is the least robust.

Of course, the choice of which composite index to use depends on several factors. In a compensatory perspective, the use of MZ is preferable to M01, for the following reasons: (1) conversion of the individual indicators into *z*-scores (the normalized variables are 'centred' with respect to the mean and have the same variance), (2) consistent correspondence grid, (3) simple interpretation of the results, (4) greater robustness (same implicit weight of the individual indicators).

In a non-compensatory perspective where we wants to assign the same weight to each indicator, JS should be excluded because the conversion into index numbers assigns different implicit weights to the normalized indicators. As for the other indices, each has its strengths and weaknesses. If the ideal unit is plausible and it is not particularly interesting to study the 'horizontal variability' of countries, we can opt for WR, as this is the most robust index. If we are more interested in interpreting the results and comparing the imbalances between the units, the MPI is more suitable as it allows the score of each country to be broken down into mean level and penalty.

Table 4.7 shows the measures of association between the five composite indices of well-being and the subjective indicator of life satisfaction of the BLI, which is a complement to the objective data to compare the quality of life in different countries. This indicator shows how people rate their life overall and is expressed on a scale from 0 to 10.

On the whole, there is a strong linear relationship between the objective well-being indices and the subjective indicator. The differences between r and ρ are also minimal, which confirms that there are no non-linear relationships and outliers that could change the correlation value. All correlation and rank correlation values are around 0.9 (− 0.9 for WR), with the exception of the values for JS ($r = 0.80$ and $\rho = 0.75$), which may be due to the different implicit weighting of this index.

Table 4.7 Relationships between composite indices of well-being and life satisfaction (Year 2023)

Composite index	Bravais-Pearson correlation coefficient	Spearman rank correlation coefficient
MZ	0.885	0.873
M01	0.892	0.898
MPI	0.888	0.886
JS	0.788	0.749
WR	− 0.901	− 0.884

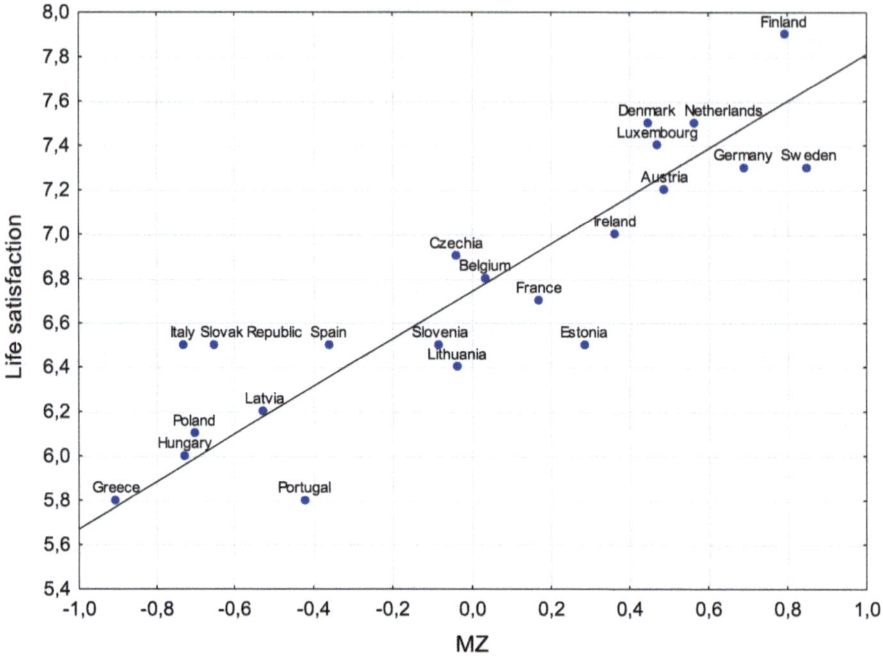

Fig. 4.4 ORLS regression between MZ and life satisfaction (Year 2023)

By using MZ as a composite index of well-being, it is possible to analyse the covariation between objective and subjective well-being in detail through symmetric regression, assuming that both variables are influenced by errors, as they are based on observational data. Figure 4.4 shows the ORLS regression line between MZ and the life satisfaction indicator.

The distribution of countries around the line confirms that in most cases objective well-being and life satisfaction go hand in hand (see e.g. Greece, Belgium and Austria). However, there are some situations where life satisfaction is higher than objective well-being (e.g. Italy) and others where life satisfaction is lower than objective well-being (e.g. Portugal).

Of course, it should be borne in mind that the objective composite index only takes into account five individual indicators and assumes that a deficit in one dimension can be offset by a surplus in another.

If we are not interested in an exact ‘estimate’ of the level of well-being for each country, we can construct the *z*-score-based performance intervals shown in Fig. 4.5 (see point 6 in Sect. 3.5.3).

The ranges of the performance intervals allow for the simultaneous representation of three types of rankings: (i) the upper bound represents the ranking in a compensatory perspective, (ii) the midpoint represents the ranking in a partially compensatory perspective, and (iii) the lower bound represents the ranking in a non-compensatory perspective.

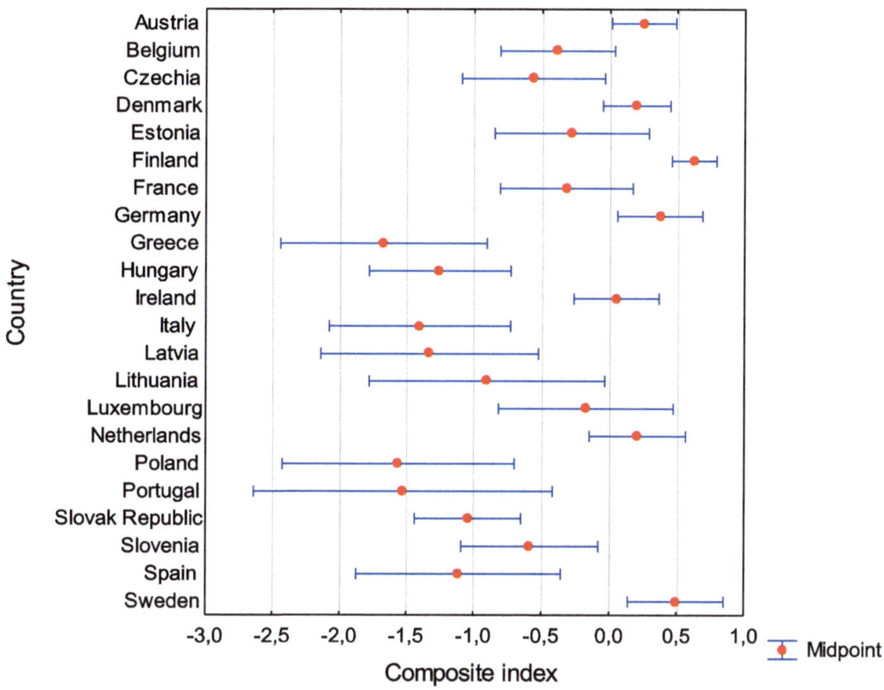

Fig. 4.5 Performance intervals of well-being (Year 2023)

Note that each country has a different range. This is due to the different horizontal variability of countries (i.e. the imbalance of normalized indicators). Therefore, one country may have a higher value than another from a non-compensatory perspective and a lower value from a compensatory perspective. In this case, the country's performance range lies within a wider range. For example, Poland has a higher non-compensatory composite index (i.e. the minimum) than Portugal, but Portugal has a higher compensatory composite index (i.e. the arithmetic mean) than Poland.

Another interesting aspect is that if the lower bound of one country is higher than the upper bound of another, its level of well-being can be considered better, regardless of the degree of compensability or substitutability of the individual indicators. Thus, Germany has a higher level of well-being than Belgium. Finally, the shorter the interval, the more balanced the country's performance. From this point of view, Finland is the country with the best balance among the normalized indicators.

4.2.2 Comparing Well-Being and GDP

Let us look at the dataset in Table 4.8, which contains nine composite indices of equitable and sustainable well-being (health, education and training, work-life balance, economic well-being, social relations, security, subjective well-being, landscape and cultural heritage, environment) and three complementary indicators for the Italian regions in 2012–2014 period. The selected indicators are as follows:

- HEA. Composite index of health (2013);
- EDU. Composite index of education and training (2014);
- QOW. Composite index of quality of work (2014);
- EMP. Normalized employment rate (2014);
- INC. Composite index of income and inequality (2014);
- HAR. Composite index of economic hardship (2014);
- REL. Composite index of social relationships (2014);
- SAF. Composite index of safety (2014);
- HOM. Normalized homicide rate (2014);
- LSI. Life satisfaction index (2014);
- LAN. Composite index for landscape and cultural heritage (2011);
- ENV. Composite index of environment (2012).

The source is the "BES Report 2015" of the Italian National Institute of Statistics (Istat) and the application is from Mazziotta and Pareto (2019).

All indicators have the same scale (Italy 2010 = 100) and a positive polarity, i.e. the higher the indicator, the greater the well-being. So, for example, the lower the homicide rate, the higher the value of HOM.

Table 4.9 shows the correlation matrix of the twelve well-being indicators and the correlation between each of them and GDP per capita (GDP) for 2014.

As we can see, most of the indicators are positively correlated among them (HEA, EDU, QOW, EMP, INC, HAR, REL, LSI and LAN), and the values are very high ($r \geq 0.700$). The composite index of environment (ENV) and homicide rate (HOM) are also positively correlated with this set of indicators, but the strength of the relationship is rather moderate for ENV ($0.450 \leq r \leq 0.700$) and weak for HOM ($0.200 \leq r \leq 0.450$). Instead, the composite indicator of safety (SAF) shows a slightly negative correlation with the other well-being indicators ($-0.250 \leq r \leq 0.200$). Regarding the correlations of the twelve well-being indicators with GDP, the highest correlation is observed with the employment rate (EMP), followed by the composite index of income and inequality (INC), the composite index of quality of work (QOW) and the composite index of health (HEA). The indicators that correlate least with GDP are the homicide rate (HOM) with $r = 0.554$ and the composite index of environment (ENV) with $r = 0.577$, whereas the composite indicator of security (SAF) is the most discordant, as it shows a negative correlation with GDP ($r = -0.221$).

Table 4.8 Indicators of well-being (Years 2012–2014)

Region	HEA	EDU	QOW	EMP	INC	HAR	REL	SAF	HOM	LSI	LAN	ENV
Piemonte	104.8	109.0	98.4	109.9	108.1	106.2	105.7	87.3	105.2	95.2	100.6	105.5
Valle d' Aosta	106.7	108.4	105.1	116.9	113.8	99.5	104.8	116.1	117.3	104.1	97.4	116.5
Liguria	110.4	113.7	98.6	106.6	105.7	96.6	100.8	86.3	104.8	94.5	95.5	103.8
Lombardia	110.7	108.7	104.3	114.8	111.4	102.1	105.3	83.7	104.0	97.5	102.7	103.9
Bolzano	125.1	124.3	108.6	126.1	118.6	109.7	129.7	95.1	113.4	118.6	115.5	124.8
Trento	113.3	125.2	106.3	117.6	113.6	107.3	120.9	105.4	117.3	109.9	119.0	112.7
Veneto	106.8	108.7	105.6	112.9	108.6	107.2	106.2	90.8	115.0	95.1	91.9	106.9
Friuli-V.G	109.2	119.1	103.5	110.9	111.4	106.4	108.4	102.9	91.4	99.1	104.0	105.9
E. Romagna	109.7	108.2	103.3	116.8	111.2	103.9	110.2	81.1	116.9	95.6	101.8	102.6
Toscana	109.3	111.1	98.8	112.4	108.8	101.8	106.6	95.4	105.6	87.9	102.7	101.5
Umbria	105.5	118.4	97.8	107.5	100.9	98.6	99.9	92.2	57.3	92.7	103.9	102.9
Marche	102.2	111.1	97.9	110.4	104.7	90.4	97.8	99.1	99.4	89.6	100.9	98.1
Lazio	101.2	110.9	90.6	103.5	95.9	99.4	99.7	82.2	97.6	88.7	94.0	100.8
Abruzzo	102.0	112.4	89.7	94.8	95.4	93.0	94.4	99.2	111.4	89.3	91.0	113.9
Molise	102.2	103.5	90.4	85.0	93.1	99.2	92.5	109.7	104.6	88.5	81.9	100.7
Campania	86.6	92.3	79.9	68.3	77.3	74.4	80.5	94.0	94.6	69.6	79.9	105.4
Puglia	93.9	95.2	82.2	73.5	88.4	81.0	83.7	83.0	98.4	82.6	92.8	103.5
Basilicata	92.5	97.5	84.6	82.6	89.0	90.9	88.2	110.7	93.2	79.1	91.9	103.5
Calabria	85.7	101.0	68.9	68.1	82.6	78.5	87.5	105.2	85.1	86.8	75.2	98.7
Sicilia	92.8	87.0	72.0	67.7	71.7	72.3	83.8	101.5	97.4	78.5	84.0	92.4
Sardegna	93.5	97.4	83.7	84.0	91.0	84.0	102.2	110.3	88.6	89.3	100.3	108.1

Table 4.9 Correlations between indicators of well-being and GDP (Years 2012–2014)

Well-being indicator	HEA	EDU	QOW	EMP	INC	HAR	REL	SAF	HOM	LSI	LAN	ENV
HEA	1.000											
EDU	0.842	1.000										
QOW	0.911	0.807	1.000									
EMP	0.917	0.850	0.963	1.000								
INC	0.906	0.841	0.963	0.969	1.000							
HAR	0.876	0.832	0.913	0.908	0.916	1.000						
REL	0.902	0.829	0.842	0.884	0.887	0.845	1.000					
SAF	− 0.232	− 0.119	− 0.201	− 0.229	− 0.172	− 0.185	− 0.084	1.000				
HOM	0.457	0.190	0.452	0.413	0.452	0.404	0.427	− 0.048	1.000			
LSI	0.871	0.826	0.791	0.821	0.858	0.785	0.927	0.020	0.418	1.000		
LAN	0.803	0.763	0.784	0.809	0.799	0.700	0.865	− 0.129	0.228	0.775	1.000	
ENV	0.559	0.540	0.538	0.494	0.555	0.478	0.639	0.178	0.428	0.696	0.532	1.000
GDP	0.889	0.748	0.889	0.928	0.899	0.834	0.873	− 0.221	0.554	0.847	0.733	0.577

These results confirm that the main well-being indicators can be 'explained' by GDP, but some of them (e.g. those related to security and environment) are almost completely independent of this measure.

An exploratory PCA was carried out to examine the overall structure of the dataset. As seen in Sect. 3.3.3, the principal components are linear combinations of the original indicators that have decreasing importance and are themselves linearly uncorrelated. This allows to describe the statistical units with a smaller number of new dimensions by maximizing the proportion of variance accounted for. In Fig. 4.6 the scree-plot (a) and the correlation circle (b) of PCA are shown. From the scree-plot examination, an elbow is evident at the second principal component. Most of the variability of the Italian regions (80.77%) can be explained by the first two principal components.

The third component accounts for 7.62% of the remaining variance, but has an eigenvalue of less than 1 (0.914) and can therefore be considered as non-significant according to the Kaiser's rule.

The correlation circle is a diagram in which each well-being indicator is represented by a point whose coordinates correspond to the two correlation coefficients with the first and second components. Each point is connected to the origin by a line and all points are bounded by a circle of unit radius whose centre is the origin.

The first principal component is highly correlated with nine out of twelve indicators (HEA, EDU, QOW, EMP, INC, HAR, REL, LSI and LAN), while the second component is highly correlated only with the composite index of safety (SAF). The normalized homicide rate (HOM) and the composite index of environment (ENV) are in an intermediate position between the two axes, as they are partially correlated with both components. Therefore, the first principal component can be interpreted as an approximate measure of well-being, while the second can be seen as a measure of security.

Figure 4.7 displays the scatterplot with ORLS regression line of the relationships between GDP per capita and the first two principal components. The correlation between GDP per capita and the first principal component is very high ($r = 0.9213$), confirming that a significant part of the information on the well-being of regions can be derived from GDP. On the other hand, it is noteworthy that the first principal component accounts for about 70% of the total variance. Therefore, GDP cannot 'capture' the remaining 30% of the information. Indeed, the second principal component, which stands for security (SAF) and partly for the environment (ENV), is completely uncorrelated with GDP per capita ($r = -0.0446$).

In Fig. 4.8 the projections of the Italian regions on the first plane of PCA are shown. The scatterplot highlights a well-known polarization between northern regions (on the right along the x-axis) and southern regions (on the left along the x-axis). The higher the first principal component, the greater the nine well-being indicators. The second factor, on the other hand, distinguishes between safer regions (at the top of the y-axis) and less safe regions (at the bottom of the y-axis). It can be concluded from this that the security of a region is almost independent of its level of well-being.

(a)

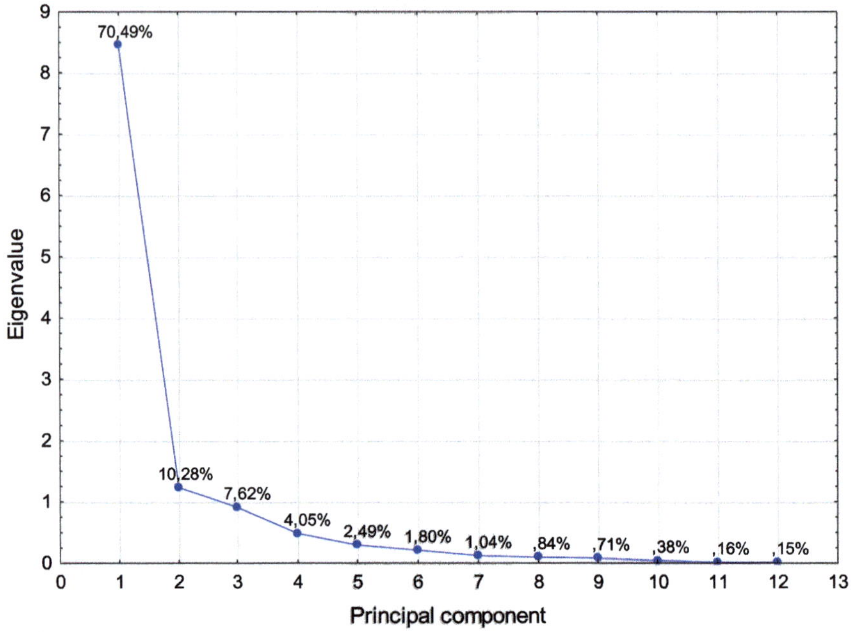

(b)

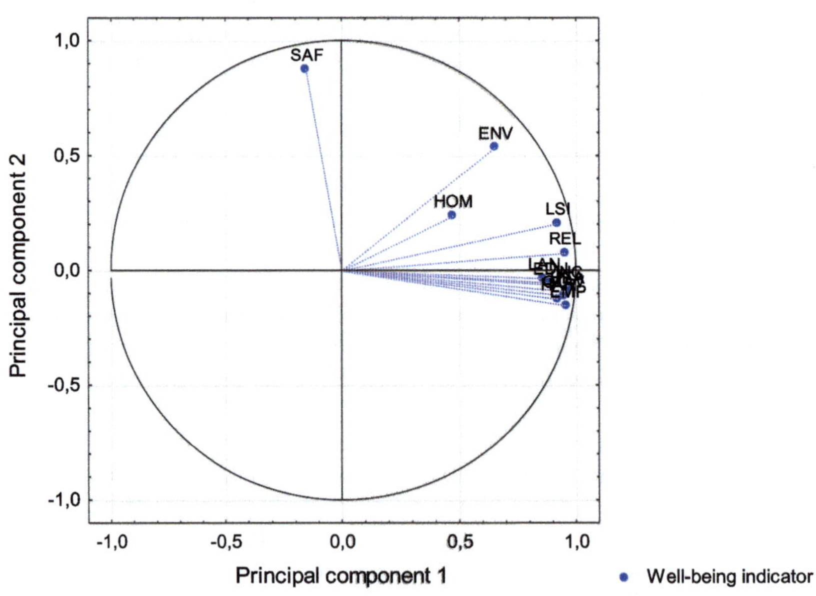

Fig. 4.6 Scree-plot and correlation circle of PCA

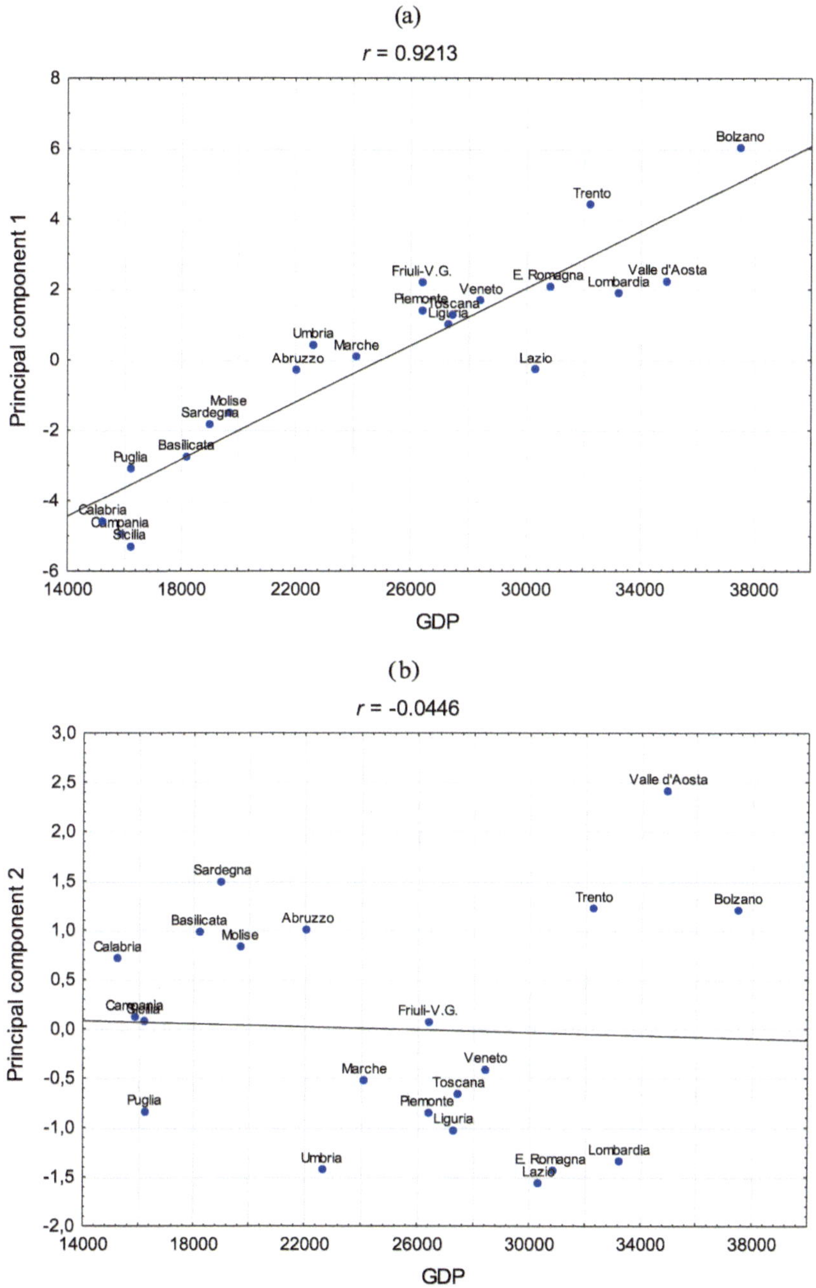

Fig. 4.7 ORLS regression between GDP and the first two principal components

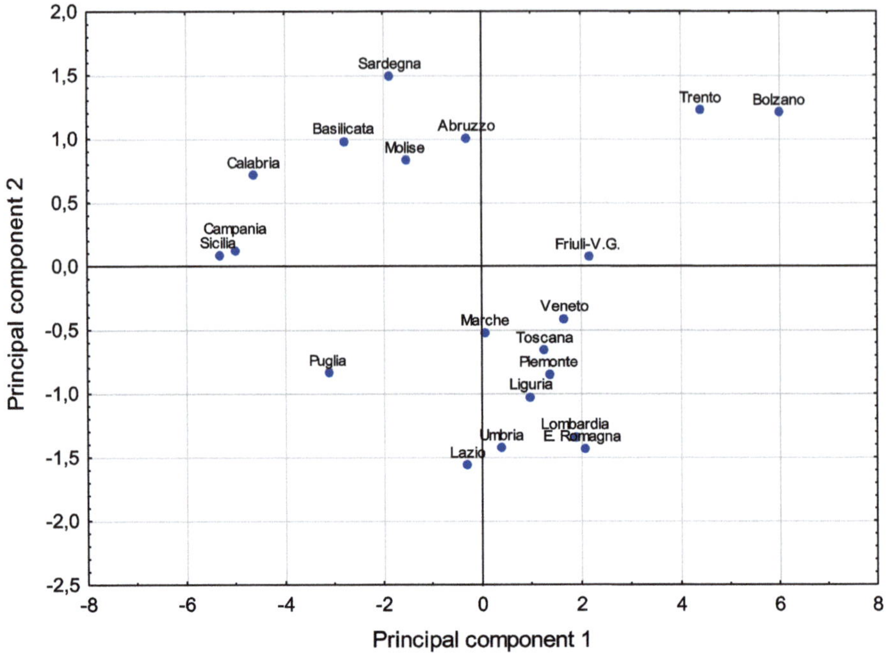

Fig. 4.8 First plane of PCA

Let us now suppose that we want to construct a composite index of general well-being. Since we have a formative measurement model, the individual indicators do not necessarily have to be correlated with each other.

Therefore, although the first principal component explains 70% of the total variance, it cannot be used as a composite index of general well-being because it represents the nine highly correlated indicators well, but gives less weight to ENV and HOM and excludes SAF. Rather, it is necessary to consider at least the first two components.

Assuming that the variance explained by each principal component reflects its importance, a PCA-based composite index can be calculated as a weighted mean of the first two principal components, with the weights given by the proportion of variance explained by each component (see point 5 in Sect. 3.5.3).

Table 4.10 shows a comparison between two possible composite indices of general well-being:

- MI. Arithmetic mean of the original indicators;
- PC. Weighted arithmetic mean of the first two principal components.

The table contains the values of the two composite indices for each region, the corresponding positions in the rankings and the difference between the two positions (difference of rank Δ).

Table 4.10 Composite indices of well-being (Years 2012–2014)

Region	MI		PC		Difference of rank (Δ)
	Value	Rank	Value	Rank	
Piemonte	0.29	9	1.11	8	1
Valle d'Aosta	0.85	3	2.25	3	0
Liguria	0.16	10	0.74	10	0
Lombardia	0.37	7	1.48	6	1
Bolzano	1.66	1	5.41	1	0
Trento	1.32	2	4.00	2	0
Veneto	0.42	6	1.41	7	1
Friuli-V.G	0.60	4	1.91	4	0
E. Romagna	0.43	5	1.63	5	0
Toscana	0.32	8	1.02	9	1
Umbria	− 0.07	13	0.19	11	2
Marche	0.01	11	0.00	12	1
Lazio	− 0.24	14	− 0.45	14	0
Abruzzo	0.01	12	− 0.13	13	1
Molise	− 0.28	15	− 1.21	15	0
Campania	− 1.34	20	− 4.31	20	0
Puglia	− 0.95	18	− 2.81	18	0
Basilicata	− 0.61	17	− 2.28	17	0
Calabria	− 1.19	19	− 3.92	19	0
Sicilia	− 1.39	21	− 4.62	21	0
Sardegna	− 0.37	16	− 1.42	16	0

As can be seen, the results are almost identical: the Spearman rank correlation between the two composite indices is 0.994 and the mean absolute difference of rank is 0.4 (i.e. each region shifts on average by 0.4 positions from one ranking to the other). More specifically, most regions do not change their position, six regions change by one position and only one region (Umbria) changes by two positions.

However, it should be noted that PC only 'captures' 80.77% of the original information.

4.2.3 Constructing Development Profiles

Let us consider the dataset in the first part of Table 4.11, which contains five individual indicators of socioeconomic development for 153 countries in the world in 2020. The dimensions examined are: health, education, wealth, work and technology. The selected indicators are as follows:

Table 4.11 Individual indicators and composite indices of socioeconomic development (Year 2020)

Country	X1	X2	X3	X4	X5	Multivariate method		Classical method		Diff. of rank (Δ)
						Index	Rank	Index	Rank	
Norway	82.4	12.9	63,633.1	14.7	96.5	2.247	1	2.126	1	0
Switzerland	83.8	13.4	68,627.8	9.8	89.7	2.187	2	1.808	3	1
Luxembourg	82.3	12.3	114,481.5	7.9	97.1	2.101	3	2.024	2	− 1
Iceland	83.0	12.8	55,874.0	9.4	99.0	2.097	4	1.665	6	2
Sweden	82.8	12.5	53,205.1	11.8	92.1	2.049	5	1.760	5	0
Singapore	83.6	11.6	97,341.5	8.0	88.2	1.963	6	1.802	4	− 2
Germany	81.3	14.2	53,815.4	5.9	89.7	1.791	7	1.354	10	3
Israel	83.0	13.0	40,161.9	9.4	81.6	1.767	8	1.411	8	0
Finland	81.9	12.8	48,621.1	7.8	88.9	1.754	9	1.375	9	0
United Kingdom	81.3	13.2	46,699.3	6.2	94.9	1.709	10	1.284	12	2
Netherlands	82.3	12.4	57,141.4	5.7	94.7	1.691	11	1.312	11	0
Ireland	82.3	12.7	86,781.4	5.1	84.5	1.690	12	1.484	7	− 5
Canada	82.4	13.4	49,031.4	5.5	91.0	1.679	13	1.261	13	0
Australia	83.4	12.7	49,756.3	5.8	86.5	1.631	14	1.246	14	0
Denmark	80.9	12.6	57,184.2	4.7	97.6	1.591	15	1.231	15	0
Austria	81.5	12.5	56,352.4	5.1	87.7	1.514	16	1.197	17	1
United States	78.9	13.4	62,682.8	4.8	87.3	1.508	17	1.214	16	− 1
Slovenia	81.3	12.7	38,688.8	7.3	79.8	1.466	18	1.148	18	0
Belgium	81.6	12.1	51,707.8	4.8	88.7	1.410	19	1.108	19	0
Japan	84.6	12.9	41,429.3	3.5	84.6	1.371	20	1.017	24	4
Estonia	78.8	13.1	36,710.4	5.6	89.4	1.365	21	1.031	23	2
Hong Kong, China (SAR)	84.9	12.3	59,892.9	1.9	89.4	1.331	22	1.069	21	− 1
Korea (Republic of)	83.0	12.2	42,661.2	3.0	95.9	1.320	23	0.988	28	5
Czechia	79.4	12.7	40,314.2	6.2	80.7	1.320	24	1.035	22	− 2
France	82.7	11.5	46,183.5	5.0	82.0	1.291	25	1.014	25	0
Malta	82.5	11.3	43,340.2	5.5	81.4	1.283	26	1.006	26	0
Poland	78.7	12.5	33,086.4	7.0	77.5	1.221	27	0.971	29	2
United Arab Emirates	78.0	12.1	67,119.1	1.9	98.5	1.141	28	0.990	27	− 1
Slovakia	77.5	12.7	32,792.6	4.9	80.7	1.043	29	0.813	30	1
Lithuania	75.9	13.1	36,974.9	4.6	79.7	1.026	30	0.802	31	1

(continued)

Table 4.11 (continued)

Country	X1	X2	X3	X4	X5	Multivariate method		Classical method		Diff. of rank (Δ)
						Index	Rank	Index	Rank	
Qatar	80.2	9.7	92,651.1	1.0	99.7	1.010	31	1.069	20	– 11
Cyprus	81.0	12.2	39,544.7	2.6	84.4	0.979	32	0.781	32	0
Spain	83.6	10.3	40,883.3	2.7	86.1	0.959	33	0.759	33	0
Italy	83.5	10.4	42,412.7	3.3	74.4	0.923	34	0.745	34	0
Greece	82.2	10.6	30,314.6	4.3	73.0	0.853	35	0.676	36	1
Latvia	75.3	13.0	30,830.1	3.3	83.6	0.827	36	0.646	37	1
Russian Federation	72.6	12.2	27,043.9	5.5	80.9	0.797	37	0.631	38	1
Brunei Darussalam	75.9	9.1	62,099.6	2.4	94.6	0.766	38	0.707	35	– 3
Portugal	82.1	9.3	34,798.0	4.0	74.7	0.765	39	0.617	40	1
Hungary	76.9	12.0	32,622.9	3.7	76.1	0.753	40	0.619	39	– 1
Croatia	78.5	11.4	28,508.5	4.3	72.7	0.745	41	0.610	41	0
Belarus	74.8	12.3	19,149.6	5.1	79.1	0.743	42	0.589	42	0
Montenegro	76.9	11.6	21,378.6	4.5	71.5	0.627	43	0.519	43	0
Bahrain	77.3	9.5	45,025.8	0.6	98.6	0.529	44	0.501	44	0
Bahamas	73.9	11.4	35,782.8	2.0	85.0	0.512	45	0.461	45	0
Georgia	73.8	13.1	15,014.3	3.9	64.0	0.464	46	0.370	49	3
Chile	80.2	10.6	24,226.2	1.4	82.3	0.434	47	0.403	46	– 1
Malaysia	76.2	10.4	28,350.6	2.2	81.2	0.389	48	0.371	48	0
Romania	76.1	11.1	29,909.0	2.5	70.7	0.387	49	0.373	47	– 2
Kuwait	75.5	7.3	49,846.1	0.8	99.6	0.383	50	0.370	50	0
Serbia	76.0	11.2	18,179.8	3.1	73.4	0.370	51	0.340	51	0
Kazakhstan	73.6	11.9	26,351.4	1.9	78.9	0.356	52	0.339	52	0
Barbados	79.2	10.6	15,638.8	1.8	81.8	0.348	53	0.330	53	0
Albania	78.6	10.1	13,961.8	3.4	71.8	0.344	54	0.318	54	0
Bulgaria	75.1	11.4	23,091.3	2.6	64.8	0.265	55	0.271	55	0
Argentina	76.7	10.9	22,033.9	1.6	74.3	0.230	56	0.265	56	0
Trinidad and Tobago	73.5	11.0	26,176.0	1.7	77.3	0.218	57	0.251	57	0
North Macedonia	75.8	9.8	16,506.1	2.3	79.2	0.183	58	0.211	59	1
Panama	78.5	10.2	31,458.7	1.3	57.9	0.148	59	0.217	58	– 1

(continued)

Table 4.11 (continued)

Country	X1	X2	X3	X4	X5	Multivariate method		Classical method		Diff. of rank (Δ)
						Index	Rank	Index	Rank	
Moldova (Republic of)	71.9	11.7	13,033.9	2.3	76.1	0.143	60	0.167	61	1
Armenia	75.1	11.3	13,653.7	2.5	64.7	0.136	61	0.175	60	− 1
Bosnia and Herzegovina	77.4	9.8	14,894.8	2.2	70.1	0.111	62	0.165	62	0
Maldives	78.9	7.0	18,914.3	3.4	63.2	0.106	63	0.103	67	4
Costa Rica	80.3	8.7	19,642.4	0.9	74.1	0.075	64	0.148	63	− 1
Turkey	77.7	8.1	28,167.4	1.6	71.0	0.074	65	0.142	64	− 1
Seychelles	73.4	10.0	29,055.7	2.3	58.8	0.054	66	0.124	65	− 1
Iran (Islamic Republic of)	76.7	10.3	14,535.9	1.5	70.0	0.025	67	0.119	66	− 1
Azerbaijan	73.0	10.6	14,403.9	1.5	79.8	0.023	68	0.103	68	0
Uruguay	77.9	8.9	21,561.1	1.2	68.3	− 0.011	69	0.092	69	0
Ukraine	72.1	11.4	12,810.3	2.0	58.9	− 0.092	70	0.002	71	1
Mauritius	75.0	9.5	22,989.3	1.6	58.6	− 0.102	71	0.020	70	− 1
Tonga	70.9	11.2	6266.9	3.9	41.2	− 0.123	72	− 0.081	75	3
Suriname	71.7	9.3	16,328.7	3.2	48.9	− 0.179	73	− 0.094	77	4
Colombia	77.3	8.5	14,730.9	1.2	62.3	− 0.213	74	− 0.063	73	− 1
Mexico	75.1	8.8	19,746.4	1.0	65.8	-0.219	75	− 0.054	72	− 3
Brazil	75.9	8.0	14,651.6	1.4	67.5	− 0.229	76	− 0.084	76	0
Dominican Republic	74.1	8.1	18,419.0	0.9	74.8	− 0.235	77	− 0.077	74	− 3
Thailand	77.2	7.9	18,463.1	1.2	56.8	− 0.258	78	− 0.108	78	0
Sri Lanka	77.0	10.6	13,078.1	1.0	34.1	− 0.285	79	− 0.163	84	5
Jamaica	74.5	9.7	9761.5	1.6	55.1	− 0.291	80	-0.126	79	-1
Samoa	73.3	10.8	6517.0	2.5	33.6	− 0.292	81	− 0.202	88	7
Mauritania	64.9	4.7	5197.0	9.9	20.8	− 0.310	82	− 0.360	98	16
Egypt	72.0	7.4	11,763.3	3.7	46.9	− 0.313	83	− 0.224	90	7
Saint Lucia	76.2	8.5	15,448.8	1.4	50.8	− 0.314	84	− 0.154	81	− 3
Palestine, State of	74.1	9.2	5661.6	1.4	64.4	− 0.353	85	− 0.160	82	− 3
Peru	76.7	9.7	12,847.9	0.4	52.5	− 0.353	86	− 0.145	80	− 6
Fiji	67.4	10.9	13,853.4	1.7	50.0	− 0.361	87	− 0.236	91	4

(continued)

Table 4.11 (continued)

Country	X1	X2	X3	X4	X5	Multivariate method		Classical method		Diff. of rank (Δ)
						Index	Rank	Index	Rank	
Mongolia	69.9	10.3	12,309.8	2.9	23.7	− 0.365	88	− 0.324	95	7
Paraguay	74.3	8.5	12,684.6	0.9	65.0	− 0.365	89	− 0.166	85	− 4
Ecuador	77.0	8.9	11,375.3	0.5	57.3	− 0.374	90	− 0.162	83	− 7
Algeria	76.9	8.0	11,349.7	0.8	59.6	− 0.380	91	− 0.185	86	− 5
Tunisia	76.7	7.2	10,755.6	1.0	64.2	− 0.387	92	− 0.205	89	− 3
Bhutan	71.8	4.1	11,345.4	4.1	48.1	− 0.412	93	− 0.412	102	9
Bolivia (Plurinational State of)	71.5	9.0	8724.5	2.5	43.8	− 0.420	94	− 0.276	94	0
Viet Nam	75.4	8.3	8041.2	0.3	70.3	− 0.433	95	− 0.198	87	− 8
Nigeria	54.7	6.7	5135.5	6.7	42.0	− 0.454	96	− 0.623	111	15
Belize	74.6	9.9	7004.9	0.8	47.1	− 0.479	97	− 0.256	92	− 5
Solomon Islands	73.0	5.7	2367.4	5.0	11.9	− 0.495	98	− 0.547	108	10
Philippines	71.2	9.4	8908.2	0.8	60.1	− 0.508	99	− 0.274	93	− 6
Kyrgyzstan	71.5	11.1	5253.1	0.8	38.0	− 0.554	100	− 0.350	97	− 3
Botswana	69.6	9.6	17,766.5	0.6	47.0	− 0.558	101	− 0.337	96	− 5
Morocco	76.7	5.6	7514.7	0.5	64.8	− 0.577	102	− 0.374	99	− 3
South Africa	64.1	10.2	12,481.8	0.8	56.2	− 0.587	103	− 0.411	101	− 2
Guatemala	74.3	6.6	8637.6	0.3	65.0	− 0.625	104	− 0.376	100	− 4
Iraq	70.6	7.3	10,881.2	1.3	49.4	− 0.660	105	− 0.448	103	− 2
Tajikistan	71.1	10.7	3379.7	0.9	22.0	− 0.664	106	− 0.502	106	0
Cabo Verde	73.0	6.3	7171.8	0.8	58.2	− 0.701	107	− 0.463	104	− 3
Indonesia	71.7	8.2	11,812.2	0.7	39.8	− 0.712	108	− 0.471	105	− 3
Guyana	69.9	8.5	9702.6	1.0	37.3	− 0.737	109	− 0.513	107	− 2
Comoros	64.3	5.1	3081.4	4.9	8.5	− 0.748	110	− 0.860	125	15
Lao People's Democratic Republic	67.9	5.3	7826.4	2.8	25.5	− 0.802	111	− 0.743	117	6
Honduras	75.3	6.6	5728.0	0.5	31.7	− 0.824	112	− 0.600	110	− 2
El Salvador	73.3	6.9	8776.0	0.5	33.8	− 0.828	113	− 0.591	109	− 4
Ghana	64.1	7.3	5412.6	2.1	39.0	− 0.837	114	− 0.693	114	0
Burundi	61.6	3.3	751.7	5.2	2.7	− 0.838	115	− 1.092	137	22

(continued)

Table 4.11 (continued)

Country	X1	X2	X3	X4	X5	Multivariate method		Classical method		Diff. of rank (Δ)
						Index	Rank	Index	Rank	
Micronesia (Federated States of)	67.9	7.8	3468.6	1.6	35.3	− 0.841	116	− 0.637	112	− 4
Nicaragua	74.5	6.9	5407.1	0.4	27.9	− 0.877	117	− 0.642	113	− 4
Namibia	63.7	7.0	9637.2	0.7	51.0	− 0.932	118	− 0.713	115	− 3
Kiribati	68.4	8.0	2275.2	1.2	14.6	− 0.961	119	− 0.805	120	1
India	69.7	6.5	6754.3	0.6	34.5	− 0.970	120	− 0.726	116	− 4
Eswatini (Kingdom of)	60.2	6.9	8688.1	1.0	47.0	− 0.978	121	− 0.831	123	2
Vanuatu	70.5	7.1	3143.6	0.8	25.7	− 0.979	122	− 0.750	118	− 4
Sao Tome and Principe	70.4	6.4	3963.8	0.8	29.9	− 0.982	123	− 0.753	119	− 4
Bangladesh	72.6	6.2	4753.7	0.5	15.0	− 1.009	124	− 0.833	124	0
Nepal	70.8	5.0	3416.8	0.7	34.0	− 1.043	125	− 0.823	122	− 3
Cambodia	69.8	5.0	4388.8	0.5	40.0	− 1.059	126	− 0.818	121	− 5
Timor-Leste	69.5	4.8	3123.1	0.8	27.5	− 1.097	127	− 0.912	126	− 1
Cameroon	59.3	6.3	3652.7	1.7	23.2	− 1.100	128	− 1.055	136	8
Côte d'Ivoire	57.8	5.3	5238.3	1.1	46.8	− 1.105	129	− 1.028	132	3
Sierra Leone	54.7	3.7	1718.4	3.0	9.0	− 1.108	130	− 1.379	151	21
Zimbabwe	61.5	8.5	2835.9	0.5	27.1	− 1.112	131	− 0.923	127	− 4
Zambia	63.9	7.2	3479.5	1.0	14.3	− 1.113	132	− 0.989	130	− 2
Senegal	67.9	3.2	3394.9	0.5	46.0	− 1.133	133	− 0.955	128	− 5
Angola	61.2	5.2	6653.9	1.5	14.3	− 1.141	134	− 1.132	139	5
Pakistan	67.3	5.2	4690.5	0.7	15.5	− 1.167	135	− 1.030	133	− 2
Togo	61.0	4.9	1595.9	1.9	12.4	− 1.168	136	− 1.179	141	5
Myanmar	67.1	5.0	5142.2	0.3	30.7	− 1.190	137	− 0.965	129	− 8
Uganda	63.4	6.2	2181.2	0.8	23.7	− 1.209	138	− 1.034	134	− 4
Mali	59.3	2.4	2327.4	2.0	13.0	− 1.210	139	− 1.376	150	11
Tanzania (United Republic of)	65.5	6.1	2660.4	0.3	25.0	− 1.227	140	− 0.999	131	− 9

(continued)

Table 4.11 (continued)

Country	X1	X2	X3	X4	X5	Multivariate method		Classical method		Diff. of rank (Δ)
						Index	Rank	Index	Rank	
Congo (Democratic Republic of the)	60.7	6.8	1097.9	0.9	8.6	– 1.240	141	– 1.178	140	– 1
Rwanda	69.0	4.4	2226.2	0.2	21.8	– 1.250	142	– 1.051	135	– 7
Niger	62.4	2.1	1219.1	1.6	5.3	– 1.264	143	– 1.411	153	10
Madagascar	67.0	6.1	1646.2	0.1	9.8	– 1.279	144	– 1.094	138	– 6
Burkina Faso	61.6	1.6	2189.6	1.3	16.0	– 1.287	145	– 1.398	152	7
Liberia	64.1	4.8	1427.8	0.8	8.0	– 1.294	146	– 1.224	143	– 3
Lesotho	54.3	6.5	2767.7	0.2	29.0	– 1.312	147	– 1.260	147	0
Papua New Guinea	64.5	4.7	4387.6	0.3	11.2	– 1.322	148	– 1.212	142	– 6
Afghanistan	64.8	3.9	2202.3	0.4	13.5	– 1.343	149	– 1.244	145	– 4
Ethiopia	66.6	2.9	2219.7	0.2	18.6	– 1.350	150	– 1.244	146	– 4
Mozambique	60.9	3.5	1280.5	0.9	10.0	– 1.357	151	– 1.376	149	– 2
Gambia	62.1	3.9	2206.9	0.5	19.8	– 1.362	152	– 1.270	148	– 4
Malawi	64.3	4.7	1059.7	0.2	13.8	– 1.376	153	– 1.231	144	– 9

- X1. Life expectancy at birth (years);
- X2. Mean years of schooling (years);
- X3. GDP per capita (PPP[3] $);
- X4. High-skill to low-skill ratio (employment ratio);
- X5. Internet users (% of population).

The source is the "Human Development Report 2020" of the United Nations Development Programme (UNDP) and the application is from Casacci and Pareto (2021).

All indicators considered have a positive polarity, so that no polarity reversal is required for normalization.

Table 4.12 shows the main summary statistics for each individual indicator. As in Sect. 4.2.1, the coefficients of variation are very different and range from a CV = 9.7% for life expectancy at birth (X1) to a CV = 99.5% for GDP per capita (X3).

No individual indicator is distributed according to the normal curve or an approximation of it. X1 and X2 have a slightly left-skewed distribution (negative

[3] At purchasing power parity.

Table 4.12 Summary statistics of the individual indicators of socioeconomic development (Year 2020)

Statistic	X1	X2	X3	X4	X5
Min	54.3	1.6	751.7	0.1	2.7
Max	84.9	14.2	114,481.5	14.7	99.7
Mean	73.3	9.0	21,890.1	2.7	56.6
Median	74.5	9.4	13,853.4	1.7	62.3
Std	7.1	3.0	21,774.3	2.6	27.7
CV %	9.7	33.7	99.5	95.5	48.9
Skewness	– 0.6	– 0.4	1.5	1.7	– 0.3
Kurtosis	– 0.3	– 0.9	2.5	3.6	– 1.2

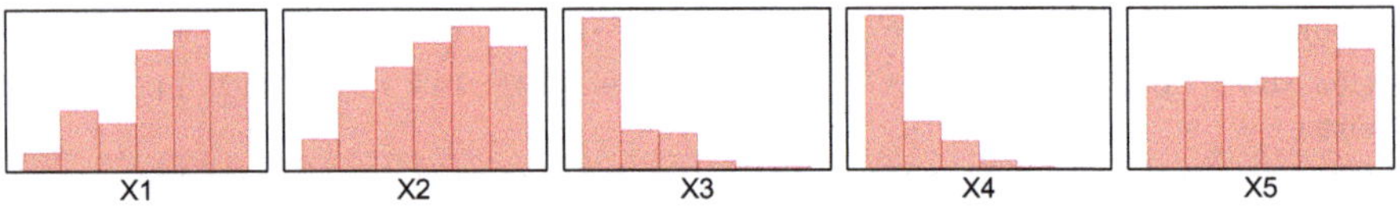

Fig. 4.9 Distributions of the individual indicators of socioeconomic development (Year 2020)

skewness), while X3 and X4 have a heavily right-skewed distribution (positive skewness). Finally, X5 has a less skewed and roughly uniform distribution. The frequency distributions of the individual indicators are shown in Fig. 4.9.

The bivariate relationships between the individual indicators are shown in Fig. 4.10 and are clearly curvilinear. This suggests the use of the multivariate method based on NLPCA to create the development profiles (see point 7 in Sect. 3.5.3).

The correlation matrices of the original and normalized indicators are shown in Table 4.13. The average correlation level between the original indicators is 0.663 and ranges from 0.444 (correlation between X1 and X4) to 0.823 (correlation between X1 and X5). These values are not particularly high, but they do not take into account the possible non-linear relationships in the data. When the indicators are normalized and the curvilinear relationships are linearized, the average correlation level becomes 0.729, with a minimum of 0.551 (correlation between X1 and X4) and a maximum of 0.912 (correlation between X3 and X5).

The correlations between the normalized indicators are higher than the correlations between the original indicators, as they are not influenced by the curvilinearity. The most pronounced curvilinearities—and thus the greatest increase in correlations with data normalization—can be observed in the relationships between indicators with negative skewness (X1-X2) and indicators with positive skewness (X3-X4). This particular effect of normalization can be easily assessed by comparing Fig. 4.10 with Fig. 4.11, which shows the bivariate relationships between the normalized indicators.

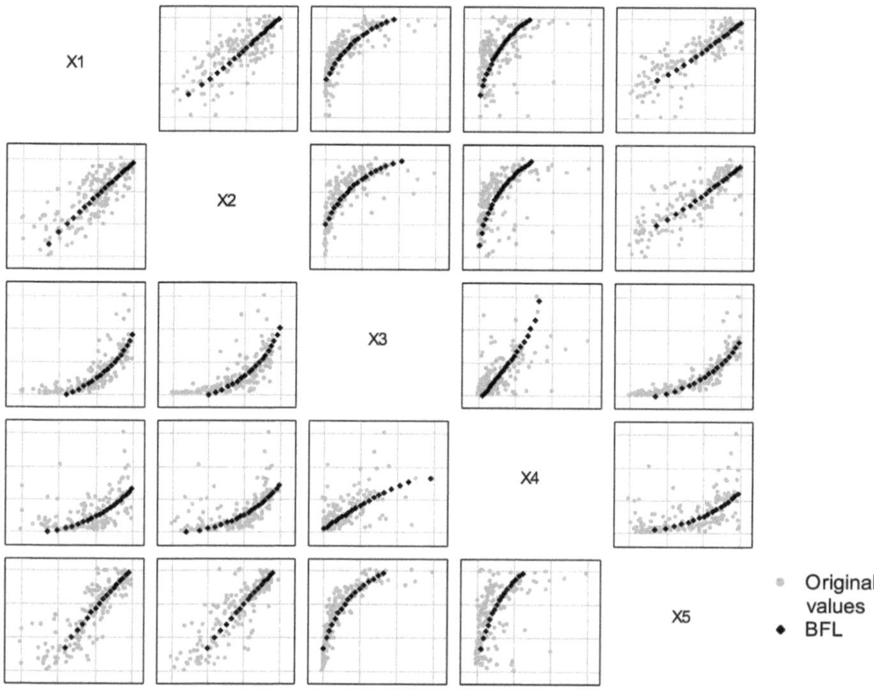

Fig. 4.10 Matrix plot of the individual indicators of socioeconomic development and BFL (Year 2020)

Table 4.13 Correlations between individual indicators of socioeconomic development (Year 2020)

Individual indicator	X1	X2	X3	X4	X5
Original values					
X1	1.000				
X2	0.753	1.000			
X3	0.721	0.685	1.000		
X4	0.444	0.527	0.600	1.000	
X5	0.823	0.811	0.795	0.467	1.000
Normalized values					
X1	1.000				
X2	0.749	1.000			
X3	0.841	0.782	1.000		
X4	0.551	0.648	0.634	1.000	
X5	0.834	0.791	0.912	0.550	1.000

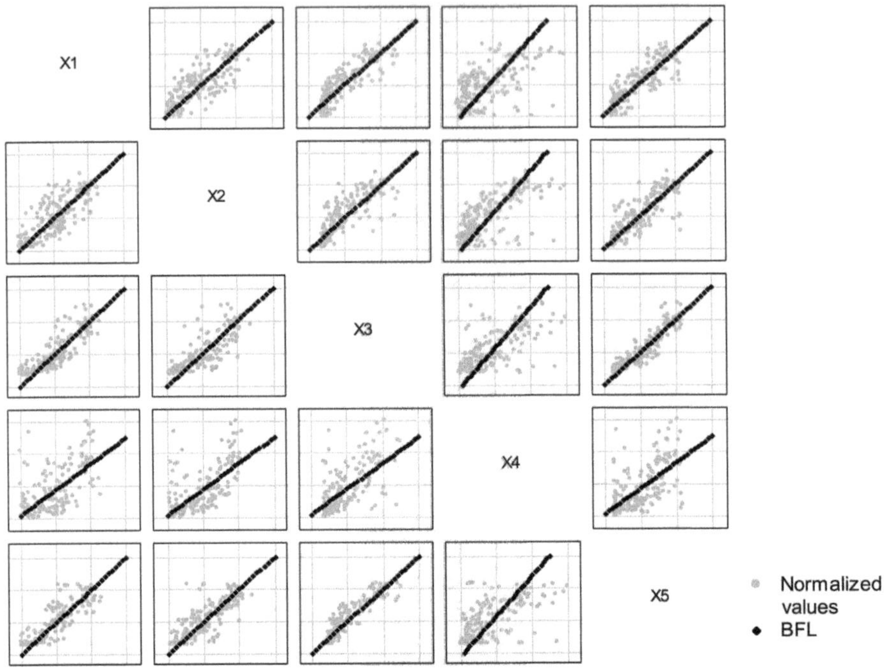

Fig. 4.11 Matrix plot of the normalized indicators of socioeconomic development and BFL (Year 2020)

Both figures also show the interpolation line, referred to in this context by the acronym BFL (best-fitting line), i.e. the first principal component of NLPCA of the original data matrix. When the BFL is projected into the space of the original values, a series of points are obtained that allow the construction of the correspondence grid. This grid is obtained by calculating the corresponding values of the original indicators for each value of the first principal component (scale).

Table 4.14 shows a comparison between the correspondence grid obtained in this way (multivariate method) and the correspondence grid based on the normalization of indicators using *z*-scores (classical method).

Note that at level 0 (in bold) of the common scale of the grid obtained with the multivariate method, the following values can be found: 74.6 years for X1 (above average), 9.6 years for X2 (above average), \$19,365 PPP for X3 (below average), 2.4 for X4 (below average) and 63.3% for X5 (above average). This means that, due to the pronounced non-linearity in the relationships between the original indicators, the set of averages is not really 'balanced' as assumed in the grid obtained with the classical method based on *z*-scores.

Moreover, in the multivariate method, the original values progress from one level to the next of the common scale by irregular (i.e. non-constant) steps. For example, the distance covered by GDP per capita (X3) on the original scale to

Table 4.14 Comparing correspondence grids (Year 2020)

Scale	Multivariate method					Classical method				
	X1	X2	X3	X4	X5	X1	X2	X3	X4	X5
2.80	–		110,941	8.1		93.3	17.4	82,858	9.8	134.3
2.60	–	–	88,088	7.5	–	91.9	16.8	78,503	9.3	128.7
2.40	–	14.0	78,934	7.0	–	90.5	16.2	74,148	8.8	123.2
2.20	84.5	13.7	70,207	6.5	–	89.0	15.6	69,793	8.3	117.6
2.00	83.7	13.4	63,217	6.1	97.6	87.6	15.0	65,439	7.8	112.1
1.80	82.9	13.1	57,299	5.6	94.7	86.2	14.4	61,084	7.3	106.5
1.60	82.1	12.7	51,937	5.2	91.8	84.8	13.8	56,729	6.8	101.0
1.40	81.3	12.4	47,002	4.8	88.8	83.3	13.2	52,374	6.3	95.5
1.20	80.5	12.0	42,399	4.5	85.6	81.9	12.6	48,019	5.7	89.9
1.00	79.6	11.7	38,080	4.1	82.3	80.5	12.0	43,664	5.2	84.4
0.80	78.7	11.3	33,990	3.7	78.9	79.1	11.4	39,310	4.7	78.8
0.60	77.7	10.9	30,092	3.4	75.3	77.6	10.8	34,955	4.2	73.3
0.40	76.8	10.5	26,371	3.0	71.6	76.2	10.2	30,600	3.7	67.7
0.20	75.7	10.0	22,801	2.7	67.6	74.8	9.6	26,245	3.2	62.2
0.00	**74.6**	**9.6**	**19,365**	**2.4**	**63.3**	**73.3**	**9.0**	**21,890**	**2.7**	**56.6**
– 0.20	73.5	9.1	16,048	2.1	58.8	71.9	8.4	17,535	2.2	51.1
– 0.40	72.3	8.5	12,840	1.8	53.8	70.5	7.8	13,180	1.7	45.6
– 0.60	71.0	8.0	9,729	1.5	48.3	69.1	7.2	8,826	1.1	40.0
– 0.80	69.6	7.4	6,709	1.2	42.0	67.6	6.6	4,471	0.6	34.5
– 1.00	68.0	6.6	3,771	0.9	34.6	66.2	5.9	116	0.1	28.9
– 1.20	66.2	5.8	909	0.7	24.7	64.8	5.3	–	–	23.4
– 1.40	64.2	4.8	–	0.4	–	63.4	4.7	–	–	17.8
– 1.60	61.5	3.2	–	0.1	–	61.9	4.1	–	–	12.3
– 1.80	56.4	–	–	–	–	60.5	3.5	–	–	6.7

move from level 0 to 0.2 is 22,801 – 19,365 = 3,437 \$ PPP, while the distance covered to move from level 1.8 to 2 is 63,217 – 57,299 = 5,918 \$ PPP. In contrast, in the classical method, the original values progress by regular (i.e. constant) steps, according to a linear relationship. In fact, the distance covered by X3 to move from one level to the next of the common scale is always equal to 26,245 – 21,890 = 4,355 \$ PPP.

Figures 4.12 and 4.13 illustrate two extreme cases of development profiles. Austria has a high level of development (mean of normalized indicators of 1.514) and all individual indicators are more or less at the same level (standard deviation of 0.14). Kuwait has a lower level of development (mean of normalized indicators of 0.383) and a very unbalanced set of individual indicators (standard deviation of 1.27). In particular, it has a deficit in X2 and X4 and a surplus in X3 and X5. If the individual indicators were normalized separately by *z*-scores (classical

Scale		X1	X2	X3	X4	X5
2.80		-	-	110,941	8.09	-
2.60		-	-	88,088	7.54	-
2.40		-	14.0	78,934	7.02	-
2.20		84.5	13.7	70,207	6.53	-
2.00		83.7	13.4	63,217	6.08	97.6
1.80		82.9	13.1	57,299	5.64	94.7
1.60	1.514	82.1	12.7	51,937	5.23	91.8
1.40	(0.14)	81.3	12.4	47,002	4.83	88.8
1.20		80.5	12.0	42,399	4.45	85.6
1.00		79.6	11.7	38,080	4.08	82.3
0.80		78.7	11.3	33,990	3.73	78.9
0.60		77.7	10.9	30,092	3.38	75.3
0.40		76.8	10.5	26,371	3.05	71.6
0.20		75.7	10.0	22,801	2.73	67.6
0.00		74.6	9.6	19,365	2.41	63.3
-0.20		73.5	9.1	16,048	2.10	58.8
-0.40		72.3	8.5	12,840	1.80	53.8
-0.60		71.0	8.0	9,729	1.51	48.3
-0.80		69.6	7.4	6,709	1.22	42.0
-1.00		68.0	6.6	3,771	0.94	34.6
-1.20		66.2	5.8	909	0.66	24.7
-1.40		64.2	4.8	-	0.39	-
-1.60		61.5	3.2	-	0.13	-
-1.80		56.4	-	-	-	-

Fig. 4.12 Development profile for Austria (Year 2020)

method), Austria would be more unbalanced (mean of normalized indicators of 1.197 and standard deviation of 0.21) and Kuwait would be less unbalanced (mean of normalized indicators of 0.370 and standard deviation of 0.93).

It is important to point out that a deviation in the profile does not necessarily mean that the country is not properly developed. The profile is an analytical tool that can help diagnose and report imbalances in development, but it is not a direct assessment tool. It tells us how is the development configuration of a country compared to others, but not necessarily how it should be. It could be used to show that countries with the same general level of development tend to develop differently in different contexts. For example, countries with few inhabitants may have a different configuration than countries with many inhabitants, and small countries may differ from large countries, but not necessarily in the same way as the former differ from the latter.

Moreover, countries in a given region or with similar economic policies or ideologies may share common characteristics (i.e. similar profiles) regardless of their overall level of development.

However, it is true that a significant deviation in a development profile may reflect a problem (e.g. a serious under or overinvestment) that should be investigated further.

The second part of Table 4.11 shows a comparison between the ranking of countries according to the composite index constructed using the multivariate method and the ranking according to the index constructed using the classical method. In

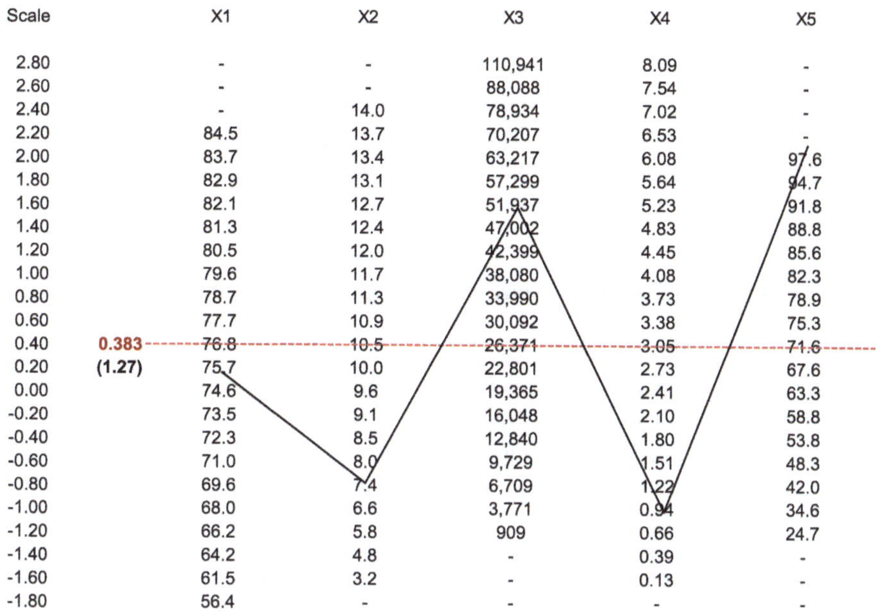

Fig. 4.13 Development profile for Kuwait (Year 2020)

both cases, the composite development index was calculated as a simple arithmetic mean of the normalized indicators. For each country, the value of the composite index of each method, the corresponding position in the ranking and the difference between the two positions (difference of rank Δ) are given.

Overall, the results are consistent and the mean absolute difference of rank between the two rankings is 3.15 (i.e. each country shifts on average by 3.15 positions from one ranking to the other). However, some countries show significant differences. For example, Burundi is ranked 115th with the multivariate method, but only 137th with the classical method (+22 positions).

4.3 Spatio-Temporal Analysis

In the first application of this section, the general government deficit/surplus indicator of the EU *Regional Competitiveness Index* (Dijkstra et al. 2011) is normalized using absolute and relative comparison methods to assess the interpretability of the temporal variations, from one year to the next.

The second application concerns the measurement of human development over time, based on the individual indicators of the *Human Development Index* (UNDP 2020). In this case, the time series of Italy from 1990 to 2021 is examined and the original index is compared with some similar composite indices obtained by normalizing the data with the constrained Min–Max method and by indicization and varying the reference base.

The third application is dedicated to the monitoring of environmental sustainability in Italy, using a set of individual indicators from the "2021 SDGs Report. Statistical Information for 2030 Agenda in Italy" (Istat 2022). For this purpose, some compensatory and non-compensatory indices are created and compared at national and geographical level from 2010 to 2019.

4.3.1 Normalizing the Government Deficit/Surplus Indicator

Let us consider the dataset in the first part of Table 4.15, which contains the "General government deficit/surplus (% of GDP)" indicator of the "Macroeconomic stability" pillar of the EU Regional Competitiveness Index (RCI) for 2016 (average 2012/14) and 2019 (average 2016/18).

The data are available at: https://ec.europa.eu/regional_policy/assets/regional-competitiveness/index.html

Supposing that this indicator is to be aggregated with the other individual indicators of the "Macroeconomic Stability" pillar, we want to examine in detail how the different normalization methods affect the variations that occur in the EU countries from one period to another.

The second part of Table 4.15 shows the values normalized by ranking, standardization and Min–Max method for relative comparisons.

As we can see, Estonia (+ 0.22%) and Luxembourg (+ 0.50%) are the only countries with a government surplus in 2016, while all the others have a deficit. Greece in particular has a deficit of − 11.23%, which lowers the overall average (mean of − 3.96% and standard deviation of 2.89%). In 2019, the situation is more balanced as 11 countries have a surplus and 17 countries have a deficit (mean of − 0.52% and standard deviation of 1.47%).

From the table we can see that for each year:

- Ranks range from 1 to 28 and have a mean of $(n + 1)/2 = (28 + 1)/2 = 14.5$.
- Standardized values (z-scores) have a mean of 0 and a standard deviation of 1, but different ranges.
- Re-scaled values with the Min–Max method range between 0 and 1, but have different standard deviations.

Let us look at Belgium, for example. From 2016 to 2019, the government deficit fell from − 3.71 to − 1.30%, but—from a ranking perspective—the country falls from 17 to 20th place, as the deficit is reduced to a greater extent in other countries. Similar results are obtained with standardization, where Belgium's normalized score drops from 0.08 (above the mean) to − 0.53 (below the mean), and with the Min–Max method, where it drops from 0.64 (vs. a mean of 0.62) to 0.38 (vs. a mean of 0.52).

Table 4.15 Government surplus/deficit indicator (% of GDP). Original and normalized values for relative comparisons (Years 2016 and 2019)

Country	Original value		Ranking		Standardization		Min–Max method	
	2016	2019	2016	2019	2016	2019	2016	2019
Austria	− 1.97	− 0.77	8.0	16.0	0.69	− 0.17	0.79	0.48
Belgium	− 3.71	− 1.30	17.0	20.0	0.08	− 0.53	0.64	0.38
Bulgaria	− 0.88	1.10	5.0	4.5	1.07	1.10	0.88	0.82
Croatia	− 6.23	0.00	23.0	11.0	− 0.79	0.35	0.43	0.62
Cyprus	− 5.71	− 0.90	21.0	17.0	− 0.61	− 0.26	0.47	0.45
Czech Republic	− 2.70	1.07	10.0	6.0	0.44	1.07	0.73	0.81
Denmark	− 2.18	0.60	9.0	9.0	0.62	0.76	0.77	0.73
Estonia	0.22	− 0.43	2.0	14.0	1.45	0.06	0.98	0.54
Finland	− 1.92	− 1.07	7.0	18.0	0.71	− 0.37	0.79	0.42
France	− 4.59	− 2.93	19.0	27.0	− 0.22	− 1.64	0.57	0.08
Germany	− 0.39	1.20	3.0	3.0	1.24	1.16	0.92	0.84
Greece	− 11.23	0.77	28.0	8.0	− 2.52	0.87	0.00	0.76
Hungary	− 3.45	− 2.00	13.0	23.0	0.18	− 1.01	0.66	0.25
Ireland	− 8.56	− 0.33	26.0	12.0	− 1.60	0.12	0.23	0.55
Italy	− 3.14	− 2.33	12.0	25.0	0.29	− 1.23	0.69	0.19
Latvia	− 1.58	− 0.50	6.0	15.0	0.82	0.01	0.82	0.52
Lithuania	− 4.63	0.47	20.0	10.0	− 0.23	0.67	0.56	0.70
Luxembourg	0.50	1.90	1.0	2.0	1.55	1.64	1.00	0.96
Malta	− 2.75	2.10	11.0	1.0	0.42	1.78	0.72	1.00
Netherlands	− 3.49	0.90	14.0	7.0	0.16	0.96	0.66	0.78
Poland	− 4.08	− 1.37	18.0	21.0	− 0.04	− 0.58	0.61	0.37
Portugal	− 6.02	− 1.83	22.0	22.0	− 0.72	− 0.89	0.44	0.28
Romania	− 3.61	− 2.80	15.0	26.0	0.12	− 1.55	0.65	0.10
Slovakia	− 3.61	− 1.23	16.0	19.0	0.12	− 0.49	0.65	0.39
Slovenia	− 8.46	− 0.40	25.0	13.0	− 1.56	0.08	0.24	0.54
Spain	− 9.09	− 3.37	27.0	28.0	− 1.78	− 1.93	0.18	0.00
Sweden	− 0.79	1.10	4.0	4.5	1.10	1.10	0.89	0.82
United Kingdom	− 6.79	− 2.10	24.0	24.0	− 0.98	− 1.07	0.38	0.23
Min	− 11.23	− 3.37	1.00	1.00	− 2.52	− 1.93	0.00	0.00
Max	0.50	2.10	28.00	28.00	1.55	1.78	1.00	1.00
Mean	− 3.96	− 0.52	14.50	14.50	0.00	0.00	0.62	0.52
Std	2.89	1.47	8.08	8.08	1.00	1.00	0.25	0.27

These results show that it can be difficult and sometimes misleading to interpret relative changes in country performance over time. Therefore, the use of normalization methods that allow absolute comparisons is recommended for spatio-temporal analyses.

Table 4.16 shows the values normalized by standardization, Min–Max method and constrained Min–Max method for absolute comparisons.

In standardization, the year 2016 was taken as the base year, and the mean and standard deviation of this year were adopted as reference values, as suggested by Tarantola (2008). The goalposts of the Min–Max method (classic and constrained) are the minimum and maximum of the indicator in the two years under consideration (− 11.23% and 2.10% respectively), while the reference value (base) for the constrained Min–Max method was set equal to 0, which represents the situation of a zero balance or equilibrium.

From the table we can see that:

- Standardized values only have a mean of 0 and a standard deviation of 1 in the base year (2016).
- Re-scaled values with the Min–Max method range between 0 and 1 from 2016 to 2019 (Greece has a value of 0 in 2016 and Malta has a value of 1 in 2019).
- Re-scaled values with the constrained Min–Max method have a clear reference value of 0 (the zero balance) in both 2016 and 2019.

In this case, all normalization methods show an increase in the score for Belgium. Indeed, the standardized value of the indicator increases from 0.08 to 0.92 (above the 2016 mean), the value normalized with the Min–Max method increases from 0.56 to 0.74 (compared to the goalposts) and the value obtained with the constrained Min–Max method increases from − 0.28 to − 0.10 (compared to the base). However, the first two increases are difficult to quantify, while the third is easy to interpret due to the presence of the base (the zero balance) and indicates that Belgium had a deficit in both 2016 and 2019, but that this was reduced and almost eliminated in 2019.

Let us now suppose that data are available for a new period (e.g. 2021) in which Belgium has a government surplus of + 2.5%.

In this case, we obtain a *z*-score of 2.24 (above the 2016 mean), a Min–Max normalized value of 1.03 (outside the range 0–1) and a constrained Min–Max normalized value of 0.19. The latter value is more informative than the standardized value (since the reference value is zero and not the 2016 mean) and, unlike the value obtained with the Min–Max method, which is no longer included in the interval [0, 1], does not require the calculation of new goalposts.

This simple application shows that in order to perform a spatio-temporal analysis, the individual indicators must be normalized using a common, easily understandable and temporally stable reference.

4.3.2 Measuring Human Development

Let us look at the dataset in Table 4.17, which contains the four individual indicators of human development for Italy in the period 1990–2021. The dimensions analysed are: health, knowledge, wealth.

Table 4.16 Government surplus/deficit indicator (% of GDP). Original and normalized values for absolute comparisons (Years 2016 and 2019)

Country	Original value		Standardization		Min–Max method		Constrained Min–Max	
	2016	2019	2016	2019	2016	2019	2016	2019
Austria	− 1.97	− 0.77	0.69	1.11	0.69	0.78	− 0.15	− 0.06
Belgium	− 3.71	− 1.30	0.08	0.92	0.56	0.74	− 0.28	− 0.10
Bulgaria	− 0.88	1.10	1.07	1.75	0.78	0.92	− 0.07	0.08
Croatia	− 6.23	0.00	− 0.79	1.37	0.37	0.84	− 0.47	0.00
Cyprus	− 5.71	− 0.90	− 0.61	1.06	0.41	0.77	− 0.43	− 0.07
Czech Republic	− 2.70	1.07	0.44	1.74	0.64	0.92	− 0.20	0.08
Denmark	− 2.18	0.60	0.62	1.58	0.68	0.89	− 0.16	0.05
Estonia	0.22	− 0.43	1.45	1.22	0.86	0.81	0.02	− 0.03
Finland	− 1.92	− 1.07	0.71	1.00	0.70	0.76	− 0.14	− 0.08
France	− 4.59	− 2.93	− 0.22	0.36	0.50	0.62	− 0.34	− 0.22
Germany	− 0.39	1.20	1.24	1.79	0.81	0.93	− 0.03	0.09
Greece	− 11.23	0.77	− 2.52	1.64	0.00	0.90	− 0.84	0.06
Hungary	− 3.45	− 2.00	0.18	0.68	0.58	0.69	− 0.26	− 0.15
Ireland	− 8.56	− 0.33	− 1.60	1.26	0.20	0.82	− 0.64	− 0.03
Italy	− 3.14	− 2.33	0.29	0.56	0.61	0.67	− 0.24	− 0.18
Latvia	− 1.58	− 0.50	0.82	1.20	0.72	0.80	− 0.12	− 0.04
Lithuania	− 4.63	0.47	− 0.23	1.53	0.49	0.88	− 0.35	0.04
Luxembourg	0.50	1.90	1.55	2.03	0.88	0.98	0.04	0.14
Malta	− 2.75	2.10	0.42	2.10	0.64	1.00	− 0.21	0.16
Netherlands	− 3.49	0.90	0.16	1.68	0.58	0.91	− 0.26	0.07
Poland	− 4.08	− 1.37	− 0.04	0.90	0.54	0.74	− 0.31	− 0.10
Portugal	− 6.02	− 1.83	− 0.72	0.74	0.39	0.70	− 0.45	− 0.14
Romania	− 3.61	− 2.80	0.12	0.40	0.57	0.63	− 0.27	− 0.21
Slovakia	− 3.61	− 1.23	0.12	0.94	0.57	0.75	− 0.27	− 0.09
Slovenia	− 8.46	− 0.40	− 1.56	1.23	0.21	0.81	− 0.64	− 0.03
Spain	− 9.09	− 3.37	− 1.78	0.21	0.16	0.59	− 0.68	− 0.25
Sweden	− 0.79	1.10	1.10	1.75	0.78	0.92	− 0.06	0.08
United Kingdom	− 6.79	− 2.10	− 0.98	0.64	0.33	0.68	− 0.51	− 0.16
Min	− 11.23	− 3.37	− 2.52	0.21	0.00	0.59	− 0.84	− 0.25
Max	0.50	2.10	1.55	2.10	0.88	1.00	0.04	0.16
Mean	− 3.96	− 0.52	0.00	1.19	0.55	0.80	− 0.30	− 0.04
Std	2.89	1.47	1.00	0.51	0.22	0.11	0.22	0.11

Table 4.17 Individual indicators, HDI and its components for Italy (Years 1990–2021)

Year	Indicator				Dimension			HDI
	X1	X2	X3	X4	Health	Knowledge	Wealth	
1990	76.99	12.82	7.44	36,208.6	0.877	0.604	0.890	0.778
1991	76.99	12.99	7.57	36,665.0	0.877	0.613	0.892	0.783
1992	77.36	13.24	7.70	36,861.8	0.883	0.624	0.893	0.789
1993	77.65	13.47	7.83	36,590.2	0.887	0.635	0.892	0.795
1994	77.87	13.92	7.96	37,314.2	0.890	0.652	0.895	0.804
1995	78.12	14.07	8.09	38,549.9	0.894	0.660	0.899	0.810
1996	78.44	14.24	8.22	39,254.0	0.899	0.670	0.902	0.816
1997	78.73	14.57	8.36	40,176.2	0.904	0.683	0.906	0.824
1998	78.81	14.77	8.49	41,091.3	0.905	0.693	0.909	0.829
1999	79.20	14.92	8.62	41,883.3	0.911	0.702	0.912	0.835
2000	79.61	15.01	8.76	42,800.1	0.917	0.709	0.915	0.841
2001	79.89	15.37	8.84	43,809.3	0.921	0.722	0.919	0.849
2002	80.11	15.59	8.95	43,948.7	0.925	0.731	0.919	0.853
2003	80.19	15.80	9.06	43,882.5	0.926	0.741	0.919	0.858
2004	81.02	15.95	9.16	44,254.6	0.939	0.748	0.920	0.865
2005	81.02	16.10	9.27	44,252.0	0.939	0.756	0.920	0.868
2006	81.42	16.28	9.39	44,658.8	0.945	0.765	0.922	0.873
2007	81.50	16.36	9.53	45,095.5	0.946	0.772	0.923	0.877
2008	81.65	16.43	9.65	43,633.3	0.948	0.778	0.918	0.878
2009	81.76	16.46	9.73	42,078.4	0.950	0.782	0.913	0.878
2010	82.14	16.47	9.82	42,150.6	0.956	0.785	0.913	0.882
2011	82.24	16.71	9.90	42,043.3	0.957	0.794	0.913	0.885
2012	82.26	16.47	10.02	40,576.3	0.958	0.792	0.907	0.883
2013	82.66	16.19	10.12	39,517.5	0.964	0.787	0.903	0.882
2014	82.91	16.11	10.22	39,490.2	0.968	0.788	0.903	0.883
2015	82.53	16.03	10.31	39,796.6	0.962	0.789	0.904	0.882
2016	83.05	15.98	10.39	41,115.4	0.970	0.790	0.909	0.887
2017	82.73	16.09	10.48	41,813.0	0.965	0.796	0.912	0.888
2018	83.18	16.18	10.57	42,450.9	0.972	0.802	0.914	0.893
2019	83.55	16.23	10.65	43,121.8	0.978	0.806	0.916	0.897
2020	82.40	16.23	10.74	39,900.6	0.960	0.809	0.905	0.889
2021	82.85	16.23	10.74	42,839.5	0.967	0.809	0.915	0.895
Inf	20.00	0.00	0.00	100.0	0.000	0.000	0.000	0.000
Sup	85.00	18.00	15.00	75,000.0	1.000	1.000	1.000	1.000
Centre	52.50	9.00	7.50	37,550.0	0.500	0.500	0.895	0.607

(continued)

Table 4.17 (continued)

Year	Indicator				Dimension			HDI
	X1	X2	X3	X4	Health	Knowledge	Wealth	
E 2021	72.90	15.40	10.60	19,351.8	0.814	0.781	0.795	0.797
M 2021	71.37	12.78	8.62	16,752.1	0.790	0.642	0.774	0.732

The selected indicators are as follows:

- X1. Life expectancy at birth (years);
- X2. Expected years of schooling (years);
- X3. Mean years of schooling (years);
- X4. Gross national income per capita (2011 PPP $).

The source is the United Nations Development Programme (UNDP) and the data can be found at: https://hdr.undp.org/data-center/documentation-and-downloads.

The table also contains the sub-indices for the three dimensions examined and the HDI. The calculation method is explained in Sect. 5.1. Below are listed the goalposts (Inf and Sup) used to normalize the individual indicators (Min–Max method), the central value (Centre) and the values for Europe and Central Asia (E 2021) and the world as a whole (M 2021) for the year 2021.

Note that the central values for health and knowledge correspond to 0.5 in the normalized scale, while the value of wealth, which is $37,550, corresponds to 0.895 for the conversion to logarithms. Figure 4.14 shows the evolution of the HDI and the three sub-indices (components) for Italy.

Table 4.18 contains similar indices (sub-indices and composite indices) calculated by normalizing the individual indicators with the constrained Min–Max method, using the difference between the goalposts (Sup-Inf) as the reference range and the 2021 values for Europe and Central Asia (base 1) and the whole world (base 2) as the base. The AMPI normalization was used to calculate the geometric mean of the three sub-indices (see Sect. 3.5.3), with the base set to 100.

Overall, the values of the composite index (CI) with base 2 (M 2021) are higher, as the values of the individual indicators at global level are lower than those of Europe and Central Asia.

In particular, the base 2 index is always greater than 100, indicating that the Italian level since 1990 is higher than the global level in 2021; while the base 1 index is greater than 100 since 1993, due to an education index that remains below the level of Europe and Central Asia in 2021 until 2008.

Note that the normalized values of each individual indicator obtained with the constrained Min–Max method are a simple linear transformation of the values provided by the classical method. Therefore, the differences between the normalized values with one base and the other are constant for each year. The differences also

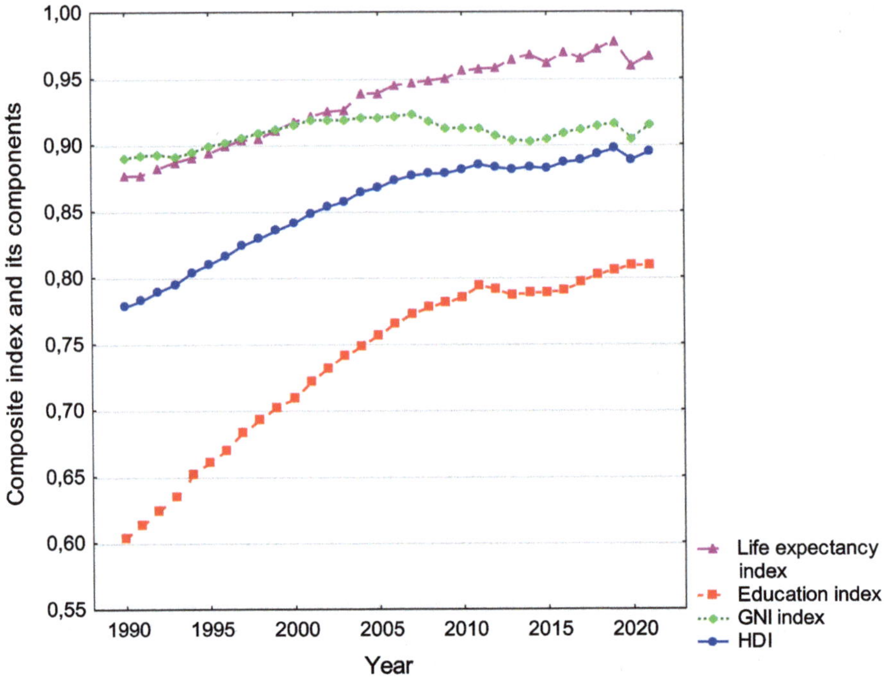

Fig. 4.14 Human development index and its components (Italy)

remain constant for the education index, which is constructed additively (i.e. linearly), while they are not preserved for the composite index, which is constructed using the multiplicative approach (i.e. non-linearly).

Figure 4.15 illustrates the evolution of the education index and its components (X2 and X3) with base 1 and base 2. As we can see, the trend of the two groups of time series is identical and is translated along the ordinate axis depending on the chosen base.

Table 4.19 shows the results obtained by converting the individual indicators into index numbers. If the original values are applied to the base without adjusting them for variability, the range of the education index increases from 76.7 (1990) to 103.3 (2021) with base 1 and from 93.3 (1990) to 125.8 (2021) with base 2. This leads to a greater implicit weighting of the sub-index in the calculation of the composite index, which in the case of base 1 is below 100 until 1996.

In this case, the differences between the normalized values of the individual indicators with one base and the other are not constant for each year, but the ratios are. However, the ratios do not remain constant, neither for the additive education index nor for the multiplicative composite index.

Figure 4.16 illustrates the evolution of the education index and its components using the two bases. The graph shows that the trend of the two sets of time series is not exactly identical (for example, the evolution of X3 with base 1 is less 'steep'

Table 4.18 Human development index and its components with normalization of the individual indicators by constrained Min–Max method for Italy (Years 1990–2021)

Year	Base 1: E 2021 = 100				Base 2: M 2021 = 100			
	Dimension			CI	Dimension			CI
	Health	Knowledge	Wealth		Health	Knowledge	Wealth	
1990	103.8	89.4	105.7	99.3	105.2	97.7	107.0	103.2
1991	103.8	89.9	105.8	99.6	105.2	98.3	107.1	103.4
1992	104.1	90.6	105.8	99.9	105.5	98.9	107.1	103.8
1993	104.4	91.3	105.8	100.3	105.8	99.6	107.1	104.1
1994	104.6	92.3	106.0	100.7	106.0	100.6	107.3	104.6
1995	104.8	92.8	106.2	101.1	106.2	101.1	107.6	104.9
1996	105.1	93.3	106.4	101.4	106.5	101.6	107.7	105.3
1997	105.4	94.1	106.6	101.9	106.8	102.5	107.9	105.7
1998	105.5	94.7	106.8	102.2	106.9	103.1	108.1	106.0
1999	105.8	95.2	107.0	102.5	107.2	103.6	108.3	106.4
2000	106.2	95.7	107.2	102.9	107.6	104.0	108.5	106.7
2001	106.4	96.4	107.4	103.3	107.9	104.8	108.7	107.1
2002	106.7	97.0	107.4	103.6	108.1	105.3	108.7	107.4
2003	106.7	97.6	107.4	103.8	108.1	105.9	108.7	107.6
2004	107.5	98.0	107.5	104.2	108.9	106.4	108.8	108.0
2005	107.5	98.5	107.5	104.4	108.9	106.8	108.8	108.2
2006	107.9	99.0	107.6	104.7	109.3	107.4	108.9	108.5
2007	107.9	99.5	107.7	104.9	109.4	107.8	109.0	108.7
2008	108.1	99.8	107.4	105.0	109.5	108.1	108.7	108.8
2009	108.2	100.0	107.0	105.0	109.6	108.4	108.3	108.8
2010	108.5	100.2	107.1	105.2	109.9	108.6	108.4	109.0
2011	108.6	100.8	107.0	105.4	110.0	109.1	108.3	109.2
2012	108.6	100.6	106.7	105.3	110.1	109.0	108.0	109.0
2013	109.0	100.4	106.5	105.2	110.4	108.7	107.8	109.0
2014	109.2	100.4	106.5	105.3	110.7	108.8	107.8	109.1
2015	108.9	100.5	106.5	105.2	110.3	108.8	107.8	109.0
2016	109.4	100.6	106.8	105.5	110.8	108.9	108.1	109.3
2017	109.1	100.9	107.0	105.6	110.5	109.2	108.3	109.3
2018	109.5	101.2	107.1	105.9	110.9	109.6	108.4	109.6
2019	109.8	101.5	107.3	106.1	111.2	109.8	108.6	109.9
2020	108.8	101.7	106.6	105.6	110.2	110.0	107.9	109.3
2021	109.2	101.7	107.2	106.0	110.6	110.0	108.5	109.7

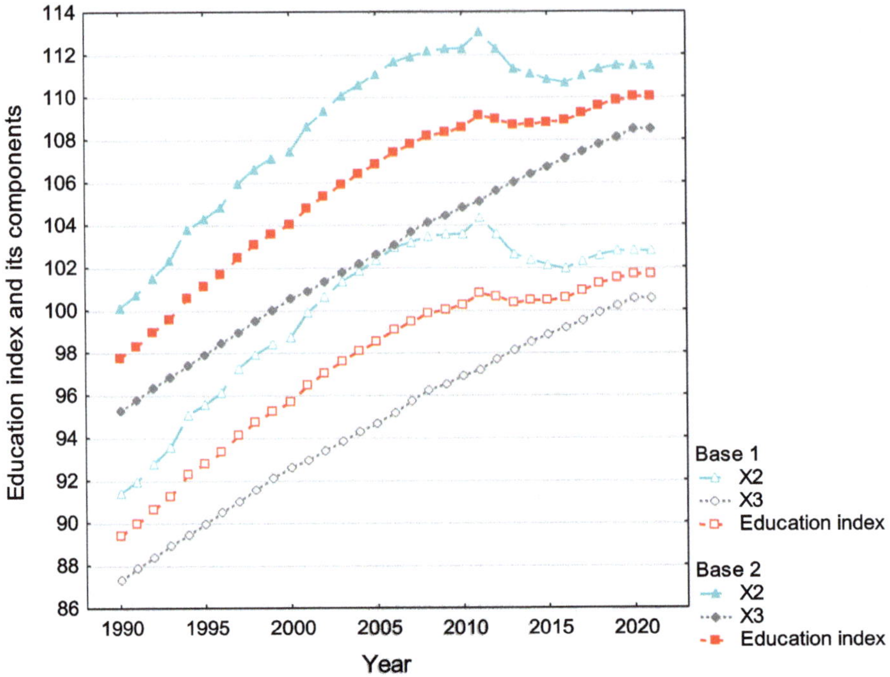

Fig. 4.15 Education index and its components (Constrained Min–Max method) (Italy)

than that with base 2) as the normalized values of X2 and X3 are proportional to the chosen base.

4.3.3 Monitoring Environmental Sustainability

Let us consider the dataset in Table 4.20, which contains four individual indicators of environmental sustainability for Italy at the geographical area level for the period 2010–2019. The targets analysed are: make cities and human settlements inclusive, safe, resilient and sustainable (goal 11); ensure access to affordable, reliable, sustainable and modern energy for all (goal 7); end hunger, achieve food security and improved nutrition, and promote sustainable agriculture (goal 2); ensure sustainable consumption and production patterns (goal 12). The selected indicators are as follows:

- X1. Air quality—PM2.5 (%);
- X2. Electricity from renewable sources (%);
- X3. Pesticides used in agriculture (Kg per hectare);
- X4. Separate collection of urban waste (%).

Table 4.19 Human development index and its components with normalization of the individual indicators by Indicization for Italy (Years 1990–2021)

Year	Base 1: E 2021 = 100				Base 2: M 2021 = 100			
	Dimension			CI	Dimension			CI
	Health	Knowledge	Wealth		Health	Knowledge	Wealth	
1990	105.6	76.7	106.3	95.2	107.9	93.3	107.9	102.8
1991	105.6	77.9	106.5	95.7	107.9	94.7	108.1	103.4
1992	106.1	79.3	106.5	96.4	108.4	96.5	108.1	104.2
1993	106.5	80.7	106.5	97.1	108.8	98.2	108.0	104.9
1994	106.8	82.7	106.7	98.1	109.1	100.7	108.2	105.9
1995	107.2	83.8	107.0	98.7	109.5	102.0	108.6	106.6
1996	107.6	85.0	107.2	99.3	109.9	103.4	108.8	107.3
1997	108.0	86.7	107.4	100.2	110.3	105.5	109.0	108.3
1998	108.1	88.0	107.6	100.8	110.4	107.1	109.2	108.9
1999	108.6	89.1	107.8	101.4	111.0	108.4	109.4	109.6
2000	109.2	90.0	108.0	102.0	111.5	109.5	109.6	110.2
2001	109.6	91.6	108.3	102.8	111.9	111.4	109.9	111.1
2002	109.9	92.8	108.3	103.4	112.3	112.9	109.9	111.7
2003	110.0	94.0	108.3	103.9	112.4	114.4	109.9	112.2
2004	111.1	95.0	108.4	104.6	113.5	115.6	110.0	113.0
2005	111.1	96.0	108.4	105.0	113.5	116.8	110.0	113.4
2006	111.7	97.1	108.5	105.6	114.1	118.2	110.1	114.1
2007	111.8	98.1	108.6	106.0	114.2	119.3	110.2	114.5
2008	112.0	98.9	108.2	106.2	114.4	120.3	109.8	114.8
2009	112.2	99.3	107.9	106.3	114.6	120.9	109.5	114.9
2010	112.7	99.8	107.9	106.7	115.1	121.4	109.5	115.2
2011	112.8	100.9	107.9	107.1	115.2	122.8	109.5	115.7
2012	112.8	100.7	107.5	106.9	115.3	122.6	109.1	115.5
2013	113.4	100.3	107.2	106.8	115.8	122.1	108.8	115.4
2014	113.7	100.5	107.2	107.0	116.2	122.3	108.8	115.6
2015	113.2	100.7	107.3	106.9	115.6	122.5	108.9	115.6
2016	113.9	100.9	107.6	107.4	116.4	122.8	109.2	116.0
2017	113.5	101.7	107.8	107.5	115.9	123.7	109.4	116.2
2018	114.1	102.4	108.0	108.0	116.6	124.6	109.6	116.7
2019	114.6	102.9	108.1	108.5	117.1	125.3	109.7	117.2
2020	113.0	103.3	107.3	107.8	115.5	125.8	108.9	116.5
2021	113.6	103.3	108.1	108.3	116.1	125.8	109.7	117.0

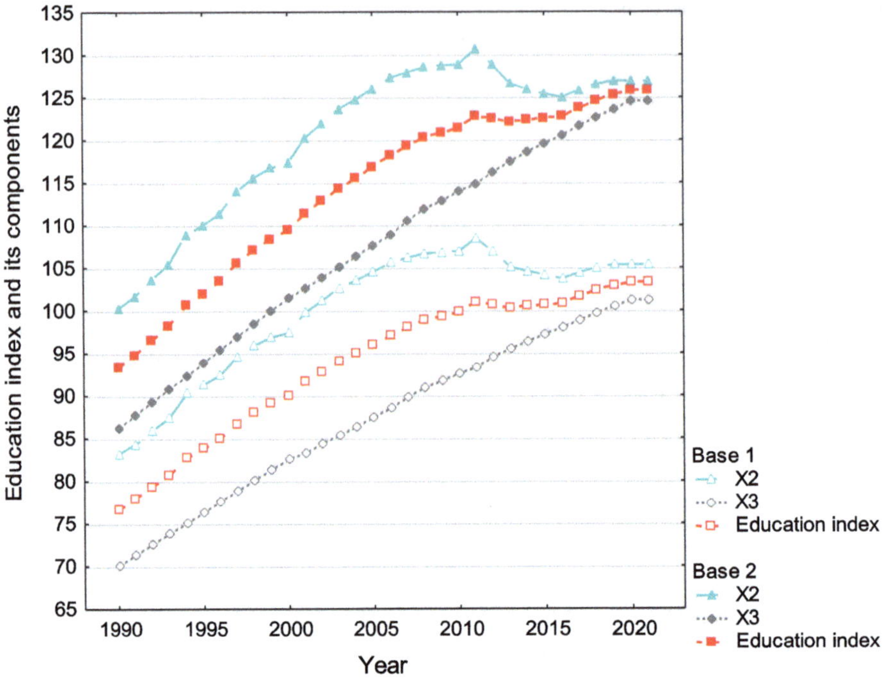

Fig. 4.16 Education index and its components (Indicization) (Italy)

The source is the Istat report on the Sustainable Development Goals (SDGs) adopted by the United Nations General Assembly with the 2030 Agenda, and the data are available at: https://www.istat.it/it/files//2018/04/Misure-statistiche-2004-2023.xlsx.

Table 4.20 also contains the minimum and maximum of the individual indicators for each year under consideration. Variables X2 and X4 have a positive polarity (e.g. the greater the share of energy from renewable sources, the greater the environmental sustainability), while X1 and X3 have a negative polarity (e.g. the greater the amount of pesticides used in agriculture, the lower the environmental sustainability).

Table 4.21 shows the values for four composite indices of environmental sustainability.

The methods used are as follows:

- M01. Arithmetic mean of indicators re-scaled to [0, 1] (Min–Max);
- MV. Arithmetic mean of indicators re-scaled with the constrained Min–Max method (Italy 2015 = 0);
- MG01. Geometric mean of indicators re-scaled to [0, 1] (Min–Max);
- AMPI. Adjusted Mazziotta-Pareto Index with negative penalty (Italy 2015 = 100).

Table 4.20 Individual indicators of environmental sustainability (Years 2010–2019)

Geographical area	2010	2011	2012	2013	2014	2015	2016	2017	2018	2019
Air quality—PM2.5 (%)										
North	97.5	97.5	95.6	95.1	94.5	96.3	94.0	94.8	94.8	91.2
Centre	92.2	93.1	92.6	90.9	82.7	87.4	80.6	75.4	80.0	74.4
South and Islands	84.6	82.4	77.9	74.7	81.7	81.2	69.9	69.5	79.8	73.4
Italy	92.9	92.2	89.4	87.7	88.5	89.7	83.8	82.4	86.8	81.9
Min	84.6	82.4	77.9	74.7	81.7	81.2	69.9	69.5	79.8	73.4
Max	97.5	97.5	95.6	95.1	94.5	96.3	94.0	94.8	94.8	91.2
Electricity from renewable sources (%)										
North	24.1	25.3	27.1	33.2	38.8	38.4	30.6	27.7	32.3	32.8
Centre	19.3	20.0	21.3	28.0	29.2	27.3	27.9	26.6	28.6	27.7
South and Islands	20.4	23.3	30.2	38.7	39.7	38.9	41.5	41.4	42.4	44.5
Italy	22.2	23.8	26.9	33.7	37.3	33.1	33.1	31.1	34.3	34.9
Min	19.3	20.0	21.3	28.0	29.2	27.3	27.9	26.6	28.6	27.7
Max	24.1	25.3	30.2	38.7	39.7	38.9	41.5	41.4	42.4	44.5
Distribution of plant protection products										
North	22.1	22.4	22.4	20.0	21.1	21.9	20.4	19.8	19.1	18.7
Centre	10.5	10.6	9.0	9.4	10.5	11.3	9.1	8.2	8.1	7.6
South and Islands	13.9	13.3	11.8	10.3	11.9	12.4	10.5	9.6	9.6	9.2
Italy	16.3	16.1	15.2	13.6	15.0	15.7	13.8	13.0	12.8	12.4
Min	10.5	10.6	9.0	9.4	10.5	11.3	9.1	8.2	8.1	7.6
Max	22.1	22.4	22.4	20.0	21.1	21.9	20.4	19.8	19.1	18.7
Separate urban waste collection										
North	49.1	51.1	52.7	54.4	56.7	58.6	64.2	66.2	67.7	69.6
Centre	27.2	30.2	33.1	36.4	40.8	43.8	48.6	51.9	54.3	58.1
South and Islands	21.2	23.9	26.5	28.8	31.3	33.6	37.6	41.9	46.1	50.6
Italy	35.3	37.7	40.0	42.3	45.2	47.5	52.6	55.5	58.2	61.3
Min	21.2	23.9	26.5	28.8	31.3	33.6	37.6	41.9	46.1	50.6
Max	49.1	51.1	52.7	54.4	56.7	58.6	64.2	66.2	67.7	69.6

M01 and MV are compensatory composite indices, while MG01 and AMPI are partially compensatory composite indices.

Since we have time series, all composite indices were constructed using normalization methods that allow absolute comparisons and take the minimum and maximum of each individual indicator from 2010 to 2019 as the goalposts (e.g. Min = 19.3 and Max = 44.5 were set for indicator X2).

The table shows that:

Table 4.21 Composite indices of environmental sustainability (Years 2010–2019)

Geographical area	2010	2011	2012	2013	2014	2015	2016	2017	2018	2019
M01										
North	0.20	0.21	0.26	0.37	0.43	0.40	0.40	0.38	0.45	0.50
Centre	0.28	0.29	0.35	0.44	0.53	0.47	0.60	0.67	0.66	0.73
South and Islands	0.27	0.34	0.49	0.64	0.57	0.57	0.75	0.79	0.73	0.84
Italy	0.25	0.28	0.37	0.49	0.51	0.46	0.57	0.59	0.60	0.67
MV (Italy 2015 = 0)										
North	– 0.26	– 0.24	– 0.20	– 0.08	– 0.03	– 0.05	– 0.06	– 0.07	– 0.01	0.05
Centre	– 0.18	– 0.16	– 0.10	– 0.01	0.08	0.02	0.15	0.21	0.21	0.27
South and Islands	– 0.19	– 0.11	0.03	0.18	0.12	0.12	0.30	0.34	0.28	0.38
Italy	– 0.21	– 0.17	– 0.09	0.03	0.05	0.00	0.11	0.13	0.14	0.21
MG01										
North	0.00	0.00	0.00	0.27	0.27	0.17	0.29	0.27	0.32	0.42
Centre	0.00	0.16	0.24	0.39	0.51	0.45	0.57	0.61	0.62	0.68
South and Islands	0.00	0.23	0.39	0.53	0.51	0.53	0.70	0.75	0.71	0.83
Italy	0.22	0.26	0.36	0.48	0.49	0.44	0.56	0.58	0.58	0.66
AMPI (Italy 2015 = 100)										
North	83.1	83.9	86.5	93.9	95.9	93.8	94.4	93.0	97.2	100.7
Centre	84.6	86.0	89.1	96.8	102.9	99.7	106.6	109.5	110.2	113.5
South and Islands	84.2	89.0	98.1	107.1	104.7	105.2	114.6	117.5	115.6	122.0
Italy	86.5	89.0	94.1	101.7	102.9	100.0	106.5	107.4	108.4	112.6

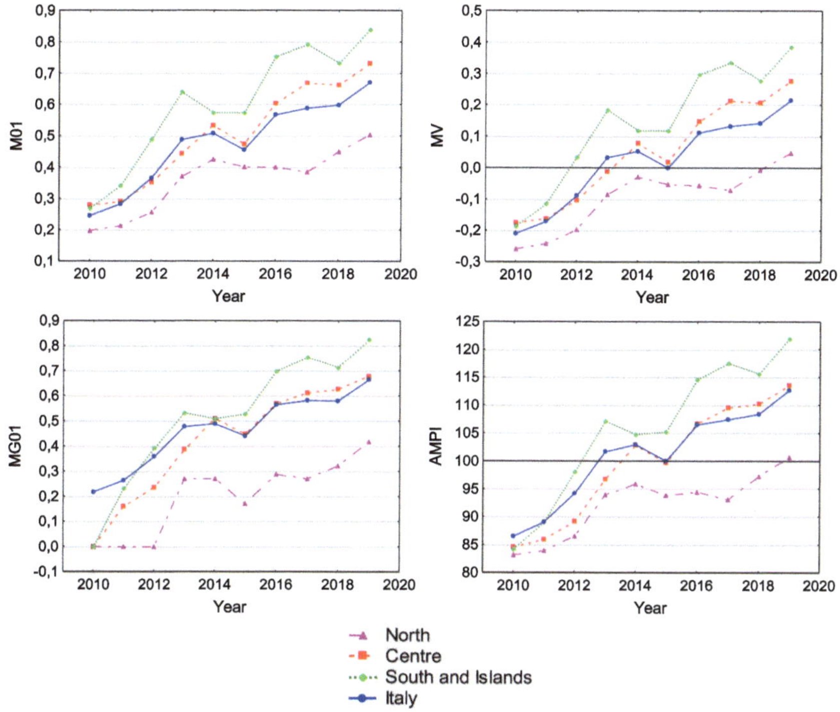

Fig. 4.17 Composite indices of environmental sustainability

- Both M01 and MG01 values are between 0 and 1. Moreover, MG01 $= 0$ if at least one individual indicator is zero[4] (see, for example, the North from 2010 to 2012), while M01 > 0 thanks to the compensation with the other values. For this reason, there is no geographical area with M01 $= 0$.
- MV values greater than zero indicate a mean value that is higher than the 2015 Italian value, while MV values less than zero indicate a lower mean value;
- AMPI values above 100 indicate a higher score (mean level + penalty) than the 2015 Italian value, while AMPI values below 100 indicate a lower score.

Figure 4.17 shows the evolution of the four composite indices over time.

Since we used an additive aggregation function (arithmetic mean), the trends of M01 and MV are identical, with the values of MV being converted so that the value for Italy in 2015 is zero.

The trends of MG01 are characterized by the presence of some zero or particularly low values due to normalized indicators that are zero or very close to zero, which significantly reduces the product of the values.

[4] Note that the geometric mean should only be used for values greater than zero.

Finally, the differences between the trends of MV and AMPI are due to the penalty effect of AMPI. Thus, in 2015, the Centre obtains a higher value than Italy with a compensatory approach (MV = 0.02) and a lower value with a partially compensatory approach (AMPI = 99.7) due to the imbalance of the individual indicators which requires the subtraction of a penalty.

References

Casacci S, Pareto A (2021) A nonlinear multivariate method for constructing profiles of country performance. Int J Multicriteria Decis Mak 8:295–312

Dijkstra L, Annoni P, Kozovska K (2011) A new regional competitiveness index: theory, methods and findings. Working Paper, No 2, DG Regional and Urban Policy

Istat (2015) BES 2015: the equitable and sustainable well-being. Istituto nazionale di statistica, Rome

Istat (2022) 2021 SDGs report. Statistical information for 2030 agenda in Italy. Istituto nazionale di statistica, Rome

Mazziotta M, Pareto A (2019) Use and misuse of PCA for measuring well-being. Soc Indic Res 142:451–476

OECD (2020) How's Life? 2020: measuring well-being. OECD Publishing, Paris

Tarantola S (2008) European Innovation Scoreboard: strategies to measure country progress over time. JRC Scientific and Technical Reports, EUR 23526 EN, Luxembourg

UNDP (2020) Human development report 2020. Statistical annex. United Nations Development Programme, New York

Some Indices Known in Literature 5

5.1 Human Development Index (HDI)

For much of the twentieth century, development was traditionally measured in terms of income or gross domestic product (GDP). Since the 1960s, however, new research approaches have been developed that contrast with the traditional approach, which essentially equates development with economic growth. In short, we have moved from a unidimensional to a multidimensional approach that takes into account not only the economic aspect but also the overall quality of life of the individual. Amartya Sen's "capability approach", which is one of the fundamental contributions in this direction, was adopted by the Human Development Report (HDR) of the United Nations Development Program (UNDP). The HDR supported a multidimensional approach to the study and measurement of development and proposed a series of composite indices to assess development and poverty. One of these indices is the *Human Development Index* (HDI), which was proposed in 1990 as a composite measure for assessing a country's development not only in terms of wealth, but also in terms of knowledge and health.

Indicators Selection

The composite index is calculated annually and measures the achievement of three key dimensions of human development: a long and healthy life, access to knowledge and a decent standard of living.

Four indicators are used to calculate the HDI: (1) "Life expectancy at birth" for the health dimension, (2) "Expected years of schooling" and (3) "Mean years of schooling" for the knowledge dimension, (4) "Gross national income (GNI) per capita" for the standard of living dimension.

This chapter was written by Sara Casacci, Ph.D. in "Social Science Methodology" and Senior Researcher at the Italian National Institute of Statistics.

M. Mazziotta and A. Pareto, *Statistics for Composite Indicators*, SxI - Springer for Innovation / SxI - Springer per l'Innovazione 18,
https://doi.org/10.1007/978-3-031-97978-1_5

Table 5.1 Individual indicators and goalpost of the HDI (Year 2018)

Dimension	Indicator	Minimum	Maximum
Health	Life expectancy at birth (years)	20	85
Knowledge	Expected years of schooling (years)	0	18
	Mean years of schooling (years)	0	15
Standard of living	Gross national income per capita (2011 PPP $)	100	75,000

Source Human development report 2021–2022 statistical update, technical notes

Normalization

The individual indicators are normalized using the Min–Max method in order to bring all values into the range [0, 1], with the minimum and maximum values given in Table 5.1 serving as goalposts.

The minimum and maximum values act as 'natural zeros' and 'aspirational targets', respectively.

The choice of 20 years as the natural zero for life expectancy at birth is based on empirical evidence that no country in the twentieth century had a lower life expectancy. The maximum life expectancy at birth is set at 85 years, a target that is considered achievable, if not already reached, in many countries.

The assumption underlying the adoption of 0 years as the minimum goalpost for the knowledge indicators is that societies can subsist without the participation of their members in formal education. The maximum value of 18 years set for the expected years of schooling corresponds to the attainment of a Master's degree in most countries. The maximum value of 15 years for mean years of schooling corresponds to the maximum value forecast for this indicator in 2025.

The minimum value for GNI per capita of $100 is justified by the significant amount of unmeasured subsistence and non-market production in some economies, which is not captured in official statistics. The maximum value is set at $75,000 per capita because some studies (Kahneman and Deaton 2010) have shown that an annual income of more than $75,000 has virtually no additional benefit for human development and well-being. Before normalization, the logarithm of GNI per capita is calculated to express the reduced importance of the income dimension as a country's GDP increases.

Aggregation

An education index $I_{Education}$ is first calculated as the arithmetic mean of the two normalized indicators for the knowledge dimension. Using the arithmetic mean allows a perfect compensation between the expected years and mean years of schooling. The HDI is then calculated using the geometric mean of the indices of the three dimensions:

$$HDI = (I_{Health} \cdot I_{Education} \cdot I_{Income})^{1/3}$$

where I_{Health} is the health index (i.e. the normalized "Life expectancy at birth" indicator) and I_{Income} is the income index (i.e. the normalized "GNI per capita" indicator).

The use of the geometric mean as an aggregation function reflects the assumption of partial compensability between the three dimensions. The decisions made for the construction of the HDI are shown in Fig. 5.1.

5.2 Italian Equitable and Sustainable Well-Being Indices (BES)

The Italian National Institute of Statistics (Istat) and the National Council for Economy and Labour (CNEL), in collaboration with representatives of the social partners and civil society, have developed a multidimensional approach for measuring *Equitable and Sustainable Well-being* (BES) in Italy. The BES project was launched with the aim of creating a conceptual and methodological framework to assess the progress of a society not only from an economic, but also from a social and environmental perspective.

Since 2013, Istat has published the annual "Report on Equitable and Sustainable Well-being", which also provides a summary by dimension using composite indices. In addition, a subset of 12 indicators from the BES has been part of the economic planning cycle since 2017. The Ministry of Economy and Finance (MEF) receives an annex to the "Economic and Financial Document", which contains the country's recent performance and an assessment of the impact of policy measures.

Indicators Selection

The definition of indicators useful for determining the state and progress of Italy is the result of an exchange with civil society and experts in the sector. The theoretical framework adopted comes from the one proposed by the OECD (Hall et al. 2010) for measuring the progress of societies by identifying relevant domains and dimensions. The framework is based on two fundamental dimensions that represent the goals to be achieved: human well-being and ecosystem well-being. The exchange process began with a survey on the importance of the dimensions of well-being, which was conducted in February 2011 as part of the Istat multipurpose survey on households "Aspects of Daily Life". On this occasion, respondents were asked to rate the importance of some dimensions of well-being by assigning a score from 0 to 10 to a list of fifteen conditions. The results of the survey, together with the evaluation of international experience, formed the basis for defining the reference framework for measuring well-being, which is based on twelve dimensions, known as domains. Some of the domains identified have a direct impact on human and environmental well-being (the so-called 'outcome domains'), while others represent functional elements for improving social and environmental well-being ('instrumental or contextual domains').

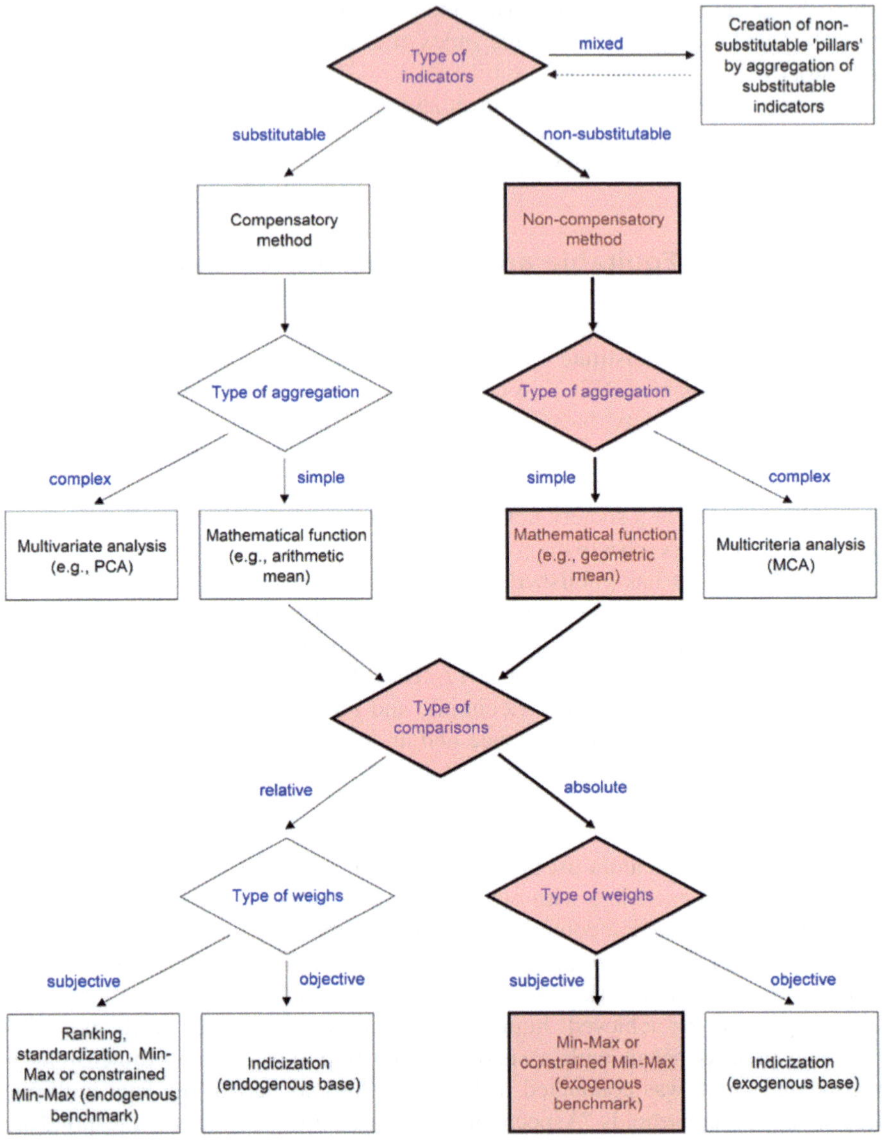

Fig. 5.1 Flowchart for constructing the HDI

Based on the domains defined by the Steering Committee, the Scientific Commission, composed of Istat researchers and professors who are experts in the various disciplines, determined the individual indicators to be used to measure well-being in each dimension. The selection was limited to indicators that are regularly available at regional level (NUTS2) to allow territorial and trend analysis of the phenomena (Istat 2015). The BES information base is constantly revised and updated.

Table 5.2 shows the individual indicators used in 2022 (Istat 2023).

Normalization and Aggregation

With regard to the twelve domains of analysis, Istat calculated a single composite index in nine cases, while two separate indices were calculated for three domains. These are the "Work and life balance" domain, for which an index for "Employment" and one for "Job quality and satisfaction"' were calculated, the "Economic well-being" domain, for which the indicators for "Income and inequality" and those for "Minimum economic conditions" were aggregated separately, and the "Safety" domain, for which "Homicide" was separated from others, less serious crimes combined into a composite measure of "Robbery".

The AMPI (Adjusted Mazziotta-Pareto Index) is used to normalize and aggregate the individual indicators for each domain. This index fulfils all the requirements deemed necessary by the Scientific Commission for the BES summaries (Istat 2015). The use of the AMPI is based on some technical and interpretative considerations (see Sect. 3.5.3). As far as normalization is concerned, most individual indicators do not have a predefined range of variation, so defining a 'closed' range (e.g. 0–1) can be misleading. For this reason, it is advantageous to normalize the indicators in an 'open' scale, where the reference level (base) is the value of Italy in 2010 (equal to 100). This enables comparisons over time and minimizes the effect caused by the different variability of the indicators. In terms of aggregation, the index is easy to interpret as the formula can be broken down into two parts: the 'mean level' (additive component) and the 'penalty' ('horizontal' variability or imbalance).

The flowchart for the BES indices is similar to that of the HDI, which is already shown in Fig. 5.1.

5.3 Canadian Index of Well-Being (CIW)

The *Canadian Index of Well-Being* (CIW) is the result of a long journey that began in 1999 when the Atkinson Foundation[1] recognized the need to create a complementary tool to GDP to measure well-being in Canada. The project involved public consultations and collaboration with numerous social indicator experts and potential users. In addition, the Canadian Research Advisory Group (CRAG) was established to identify the dimensions and develop the methodological aspects related to the creation of the index. In 2011, the project was taken over by the University of Waterloo and the first version of the composite index of well-being

[1] The Atkinson Foundation is a Canadian non-governmental charitable organization established in 1942.

Table 5.2 Individual indicators of the BES (Year 2022)

Domain	Indicator
Health	Life expectancy at birth
	Healthy life expectancy at birth
	Mental health index (SF36)
	Avoidable mortality (age 0–74)
	Infant mortality rate
	Road accidents mortality rate (15–34 years old)
	Age-standardized cancer mortality rate (20–64 years old)
	Age-standardized mortality rate for dementia and nervous system diseases (65 years and over)
	Multimorbidity and severe limitations (75 years and over)
	Life expectancy without activity limitations at 65 years of age
	Overweight or obesity (standardized rates)
	Smoking (standardized rates)
	Alcohol consumption (standardized rates)
	Sedentariness (standardized rates)
	Adequate nutrition (standardized rates)
Education and training	Children aged 0–2 years enrolled in nursery school
	Participation in the school system of children aged 4–5
	People with at least upper secondary education level (25–64 years old)
	People having completed tertiary education (25–34 years old)
	First-time entry rate to university by cohort of upper secondary graduates
	Early leavers from education and training
	People not in education, employment, or training (NEET)
	Participation in life-long learning
	Inadequate level of literacy (students in grade 8)
	Inadequate level of numeracy (students in grade 8)
	Individuals with basic or above basic overall digital skills
	STEM graduates
	Cultural participation outside the home
	Reading books and newspapers
	Use of libraries
Work and life balance	Employment rate (20–64 years old)
	Non-participation rate

(continued)

Table 5.2 (continued)

Domain	Indicator
	Transition rate (12 months time-distance) from temporary to permanent employment
	Share of employed persons with temporary jobs for at least 5 years
	Share of employees with below 2/3 of median hourly earnings
	Share of over-qualified employed persons
	Incidence rate of fatal occupational injuries or injuries leading to permanent disability
	Employed persons not in regular occupation
	Ratio of employment rate for women aged 25–49 with at least one child aged 0–5 to the employment rate of women 25–49 years without children
	Share of employed people aged 15–64 years working over 60 h per week (including paid work and household work)
	Household workload inequality index
	Job satisfaction
	Share of employed persons who feel their work unsecure
	Involuntary part time
	Employed persons working from home
Economic well-being	Gross disposable income per capita
	Disposable income inequality
	People at risk of poverty
	Per capita net wealth
	Absolute poverty (incidence)
	Severe material and social deprivation rate—Europe 2030
	Severe housing deprivation
	Great difficulty in making ends meet
	Very low work intensity—Europe 2030
	Housing cost overburden rate
	Economic situation of the household
Social relationships	Satisfaction with family relations
	Satisfaction with friends relations
	People to rely on
	Social participation
	Civic and political participation
	Voluntary activity

(continued)

Table 5.2 (continued)

Domain	Indicator
	Association funding
	Nonprofit organizations
	Generalized trust
Politics and institutions	Voter turnout
	Trust in the parliament
	Trust in judicial system
	Trust in political parties
	Trust in police and fire brigade
	Women and political representation in national parliament
	Women and political representation at regional level
	Women in decision-making bodies
	Women in the boards of companies listed in stock exchange
	Mean age of members of national parliament
	Length of civil proceedings
	Prison density
Safety	Intentional homicide rate
	Burglary
	Pick-pocketing
	Robbery
	Physical violence on women
	Sexual violence on women
	Intimate partnership violence
	Worries of being victim of a sexual violence
	Perception of safety walking alone in the dark
	Concrete fear of crime
	Social decay (or incivilities)
	Perception of crime risk
Subjective well-being	Life satisfaction
	Leisure time satisfaction
	Positive judgement of future perspectives
	Negative judgement of future perspectives
Landscape and cultural heritage	Current expenditure of municipalities on culture
	Density and importance of museums' heritage
	Illegal building rate
	Erosion of farmland from urban sprawl

(continued)

Table 5.2 (continued)

Domain	Indicator
	Erosion of farmland from abandonment
	Pressures of mining and quarrying activities
	Impact of forest fires
	Spread of agritourism farms
	Density of historic parks/gardens
	Dissatisfaction with the landscape of the place of living
	Concern about landscape deterioration
Environment	Air quality—PM2.5
	Emissions of CO_2 and other greenhouse gases
	Warm spell duration index
	Extreme precipitation events
	Consecutive dry days
	Population at risk of landslides
	Population at risk of flood
	Water losses in urban supply system
	Sewage treatment
	Protected natural areas
	Coastal bathing waters
	Urban green
	Soil sealing from artificial land cover
	Domestic material consumption
	Municipal waste generated
	Landfill of waste
	Contaminated sites
	Electricity from renewable sources
	Concern for climate change and greenhouse effect
	Satisfaction for the environment
	Concern for biodiversity loss
Innovation, research and creativity	R&D intensity
	Patent propensity
	Gross Fixed Capital Formation (GFCF) of intellectual property products
	Innovation rate of the national productive system
	Impact of knowledge workers out of employment
	Cultural employment (% of total employment)
	Brain circulation (Italians, 25–39 years old)

(continued)

Table 5.2 (continued)

Domain	Indicator
	Regular internet users
	Availability of at least one computer and Internet connection in the household
	Municipalities with online services for families
	Enterprises with at least 10 persons employed with web sales to end customers
Quality of services	Beds in the residential social-healthcare and social-welfare facilities
	Integrated home assistance service
	Composite index of service accessibility
	Irregularities in water supply
	Irregularities in electric power distribution
	Seat-Km of public transport networks
	Satisfaction with public transport services
	Frequent users of public transport
	Overall Fixed Very High Capacity Network (VHCN) coverage
	Separate collection service for municipal waste
	Hospital beds in high-care wards
	Hospital patient emigration to a different region
	Unmet need for medical examination
	General practitioners with a number of patients above the maximum threshold
	Physicians
	Nurses and midwives

was published. It is a single value that represents the percentage change (increase or decrease) in quality of life compared to the base year 1994.[2]

Indicators Selection

The CIW aims to assess well-being as the presence of the highest possible quality of life, focusing on the following aspects: good living standards, robust health, a sustainable environment, vital communities, an educated population, balanced time use, high levels of democratic participation, and access to and participation in leisure and culture (Canadian Index of Wellbeing 2012). Eight domains were identified for its definition, each of which comprises eight individual indicators, which are listed in Table 5.3.

[2] The choice of 1994 as the base year is due to the fact that most Canadian health statistics come from the National Population Health Survey, which was introduced in that year.

Table 5.3 Individual indicators of the CIW (Year 2014)

Domain	Indicator
Community vitality	Percentage of population reporting unpaid participation in any activities on behalf of a group or organization in past year
	Percentage of population with 5 or more close friends
	Crime severity index
	Experience of discrimination
	Percentage of population who feel safe walking alone after dark
	Percentage of population who believe that most or many people can be trusted
	Percentage of population who provide unpaid help to others on their own
	Percentage of population reporting a somewhat or very strong sense of belonging to community
Democratic engagement	Percentage of voter turnout at federal elections
	Percentage who volunteer for a law, advocacy, or political group
	Old and young voters' gap
	Percentage reporting they are very/fairly satisfied with way democracy works in Canada
	Percentage with a great deal or quite a lot of confidence in federal parliament
	Ratio of registered to eligible voters
	Percentage of women in parliament
	Members of parliament send communications to householders
Education	Percentage of children aged 0–5 years for whom there is a regulated centre-based child care space
	Amount of time spent in talk-based activities with children age 0–14 years
	Ratio of students to educators in public schools
	Average expenditure per public school student ($2013)
	Percentage of population age 25 and older participating in education-related activities
	Average annual Canadian undergraduate tuition fees ($2015)
	Percentage of Canadians 20–24 years of age in labour force completing high school
	Percentage of 25–64 year olds in population with a university degree
Environment	Ground level ozone (population weighted in parts per billion)
	Absolute GHG emissions (megatonnes of CO_2 per year)
	Primary energy production (petajoules)

(continued)

Table 5.3 (continued)

Domain	Indicator
	Residential energy use
	Total farm land (hectares)
	Viable metal reserves index
	Ecological footprint
	Annual water yield in Southern Canada (km^3)
Health	Percentage self-rated health as excellent or very good
	Percentage who rate their mental health as very good or excellent
	Percentage with self-reported diabetes
	Life expectancy at birth, years
	Percentage of daily or occasional smokers among teens aged 12–19 years
	Percentage of Canadians with a regular medical doctor
	Percentage of population getting influenza immunization
	Percentage of population with an absence of health limitations
Leisure and culture	Average percentage of time spent on the previous day in social leisure activities
	Average percentage of time spent on the previous day in arts and culture activities
	Average number of hours in the past year volunteering for culture/recreation orgs
	Average monthly frequency of participation in physical activity lasting over 15 min
	Average attendance per performance in past year at all performing arts performances
	Average visitation per site in past year to all National Parks and National Historic Sites
	Average number of nights away/vacation trip in past year to destinations 80 km from home
	Expenditures in past year on all culture/recreation as percentage of total household expenditures
Living standards	Gini coefficient
	After tax median income of economic families (2013$)
	Percentage of persons in low income
	Percentage of households that are moderately or severely food insecure
	Percentage labour force with long-term unemployment
	Percentage of labour force employed
	CIBC index of employment quality (1994 QI = 100)

(continued)

Table 5.3 (continued)

Domain	Indicator
	Shelter Consumption Affordability Ratio (SCAR)
Time use	Percentage of Canadians 25–64 years of age working over 50 h per week at main job
	Percentage of 15–64 year olds reporting high levels of time pressure
	Average daily amount of time with friends (minutes per day)
	Percentage of labour force working under 30 h per week, not by choice
	Percentage of Canadians who report 7–9 h of good quality essential sleep
	Mean workday commute time for individuals working for pay
	Percentage of individuals working for pay with flexible work hours
	Percentage of labour force with regular, weekday work hours

The indicators representing each domain were selected by the research groups based on a number of criteria including validity, quality, reliability and accessibility (Michalos et al. 2011).

Normalization
Normalization is performed by index number transformation. In particular, the value assumed by each individual indicator is divided by the value recorded in the base year (1994) and then multiplied by 100. For the indicators with negative polarity, the reciprocal value is calculated according to (3.2). Therefore, a non-linear transformation is carried out for the indicators with negative polarity.

Aggregation
The technical problem of constructing a composite index is solved in a compensatory manner by calculating an average of the percentage ratios.

All indicators are weighted equally and an arithmetic mean is used to construct a composite index (or better a sub-index) for each domain. Finally, the CIW is the simple arithmetic mean of the eight sub-indices, i.e. for a given year, the mean values for the eight domains are added together and then divided by eight.

The decisions made to form the CIW are summarized in Fig. 5.2.

5.4 Regional Competitiveness Index (RCI)

The EU *Regional Competitiveness Index* (RCI) measures the most important factors of competitiveness in the regions of the European Union. According to the World Economic Forum (WEF), competitiveness at the national level is the set of

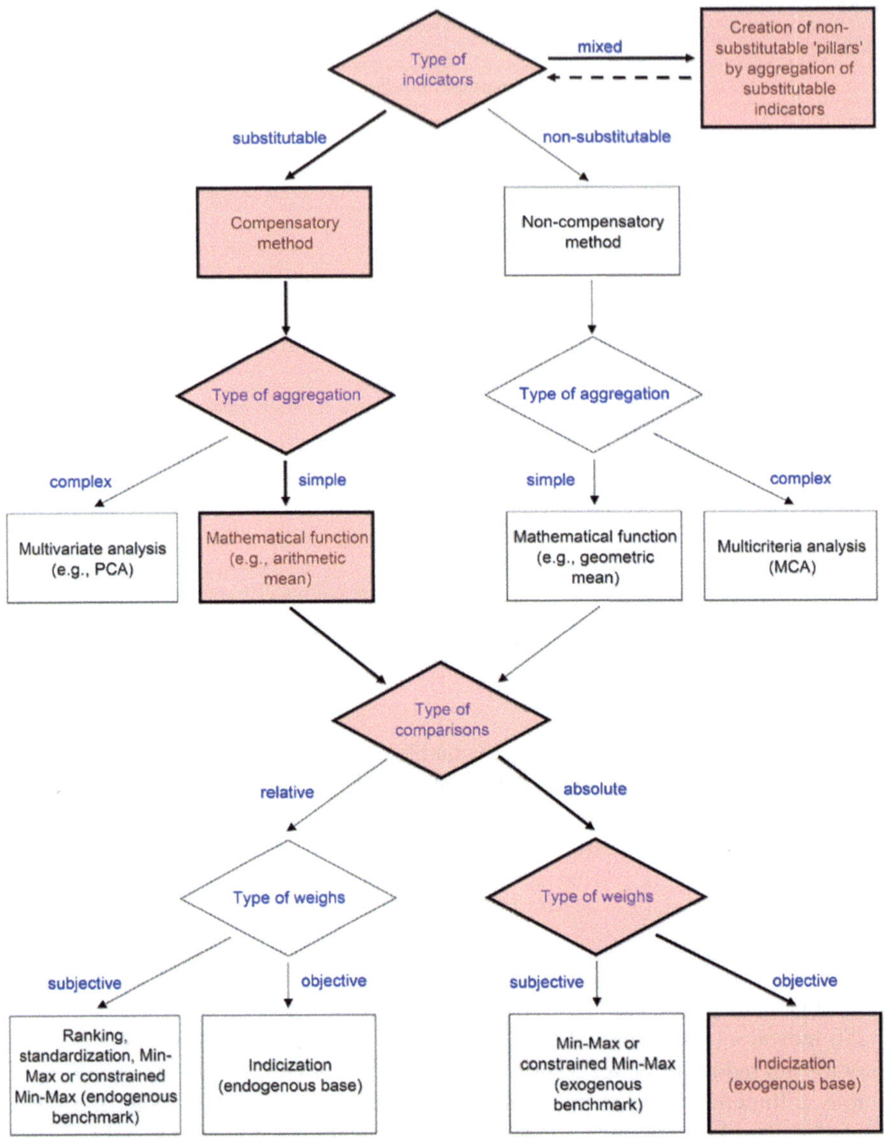

Fig. 5.2 Flowchart for constructing the CIW

institutions, policies and factors that determine a country's level of productivity (Schwab 2012; Schwab and Porter 2007). The definition of regional competitiveness chosen for the RCI is slightly different, as it includes both the perspective of businesses and residents: "*Regional competitiveness is the ability of a region to offer an attractive and sustainable environment for firms and residents to live and work*" (Dijkstra et al. 2011).

Indicators Selection
The index is composed of three sub-indices—"Basic", "Efficiency" and "Innovation"—and eleven pillars that describe various aspects of competitiveness (Annoni and Kozovska 2010).

The "Basic" sub-index refers to the most important basic factors of all types of economies. It identifies the most important aspects required for the development of regional competitiveness and comprises five pillars: (1) institutions, (2) macroeconomic stability, (3) infrastructure, (4) health and (5) basic education. The "efficiency" sub-index includes three pillars: (6) higher education, training and lifelong learning, (7) labour market efficiency and (8) market size. Finally, the "Innovation" sub-index consists of the three pillars that represent the drivers for improvement in the most advanced stage of economic development: (9) technological readiness, (10) business sophistication and (11) innovation. Overall, the RCI 2022 is based on 68 individual indicators, which are listed in Table 5.4.

Normalization
Until 2022, normalization was carried out using z-scores. Since 2022, the Min–Max method has been used to convert the values to the [0, 100] scale, with the minimum and maximum of the observed values serving as goalposts. The introduction of the new method was justified on the grounds of better interpretability and comprehensibility for a less tech-savvy audience.

Aggregation
The individual indicators within each pillar are aggregated by a simple arithmetic mean. The sub-indices are also simple arithmetic means of the underlying pillars. A weighted arithmetic mean of the three sub-indices is used to calculate the overall RCI. The weighting is distributed to the various countries according to their level of development, as shown in Table 5.5.

The scheme for constructing the RCI is shown in Fig. 5.3.

5.5 Environmental Performance Index (EPI)

The *Environmental Performance Index* (EPI) was developed in 2002 by the Yale Centre for Environmental Law & Policy (Yale University) and the Centre for International Earth Science Information Network (Columbia University) in collaboration with the World Economic Forum. The EPI is designed to provide a quantitative assessment of the environmental performance of a country's policies, based on a set of indicators covering a wide range of environmental issues, such as air and water quality, biodiversity, climate change and resource management. The aim is to provide a comprehensive picture of a country's efforts and success in tackling environmental challenges (Wolf et al. 2022).

Table 5.4 Individual indicators of the RCI (Year 2022)

Pillar	Indicator
Institutions	Corruption
	Quality and accountability
	Impartiality
	Individuals who used the internet for interaction with public authorities
	Presence of corruption in the national public institutions in the country
	Presence of corruption in the local or regional public institutions in the country
	Ease of doing business
	Property rights
	Intellectual property protection
	Efficiency of legal framework in settling disputes
	Efficiency of legal framework in challenging regulations
	Organized crime
	Reliability of police services
	Judicial independence
Macroeconomic stability	General government deficit/surplus
	National savings
	Government bond yields
	Government debt
	Net international investment position
Infrastructures	Road transport performance
	Rail transport performance
	Accessibility to passenger flights
Health	Road fatalities
	Healthy life expectancy
	Infant mortality
	Cancer disease death rate
	Heart disease death rate
	Suicide death rate
Basic education	Low achievement in reading (15-year-olds)
	Low achievement in maths (15-year-olds)
	Low achievement in science (15-year-olds)
Higher education and lifelong learning	Higher educational attainment
	Lifelong learning

(continued)

Table 5.4 (continued)

Pillar	Indicator
	Early school leavers
	University accessibility
	Lower secondary completion only
Labour market efficiency	Employment rate (excluding agriculture)
	Long-term unemployment
	Unemployment rate
	Labour productivity
	Gender balance unemployment
	Gender balance employment
	NEET
	Labour market slack
	Temporary employment
Market size	Disposable income per capita
	Potential market size expressed in GDP
	Potential market size expressed in population
Technological readiness	Households with broadband access
	Individuals buying over internet the last year
	Access to high-speed broadband
	Individuals with above basic overall digital skills
	Enterprises having received orders online (at least 1%)
	Enterprises with fixed broadband access
Business sophistication	Employment (K-N sectors)
	Gross value added (GVA) (K-N sectors)
	Innovative SMEs collaborating with others
	Marketing or organizational innovators
Innovation	Total patent applications
	Core creative class employment
	Knowledge workers
	Scientific publications
	Total intramural research and development expenditure
	Human Resources in Science and Technology
	Employment in technology and knowledge intensive sectors
	Trademark applications

(continued)

Table 5.4 (continued)

Pillar	Indicator
	Design applications
	Sales of new-to-market and new-to-firms innovation

Table 5.5 Weighting scheme for the three RCI sub-indices, by level of development (Year 2022)

Stage of development	Sub-index weights		
	Basic	Efficiency	Innovation
Stage 1: GDP < 75	0.3	0.5	0.2
Stage 2: GDP [75–100]	0.25	0.5	0.25
Stage 3: GDP > 100	0.2	0.5	0.3

Source Dijkstra et al. (2023)

Indicators Selection

The index comprises 40 indicators, which are divided into eleven issue categories and three policy objectives. They provide information at national level on how close the countries are to the defined environmental policy targets. The individual indicators for 2022 are listed in Table 5.6.

Normalization and Aggregation

Firstly, some indicators are logarithmically transformed. Secondly, the indicators are re-scaled to [0, 100] using the Min–Max method, so that each country is positioned in relation to the 'worst' and best' performance target, which correspond to scores of 0 and 100 respectively. The best performance targets are selected based on the following hierarchy of criteria (Wendling et al. 2020):

- international agreements, treaties or institutions;
- recommendation of expert judgement;
- 95th or 99th percentile of the indicator (depending on the distribution and the type of indicator).

The determination of the worst performance targets follows a similar logic, although the first two criteria are rarely available.

The EPI is determined by successive levels of aggregation. Normalized indicators are aggregated into issue category indices, issue category indices into policy objective sub-indices and policy objective sub-indices into the EPI. A weighted arithmetic mean is used at each aggregation level. Each indicator is assigned a different weight. The weights reflect a combination of aspects determined by subjective judgement, data quality, and analysis of global trends.

The scheme for creating the EPI is the same of the RCI and is shown in Fig. 5.3.

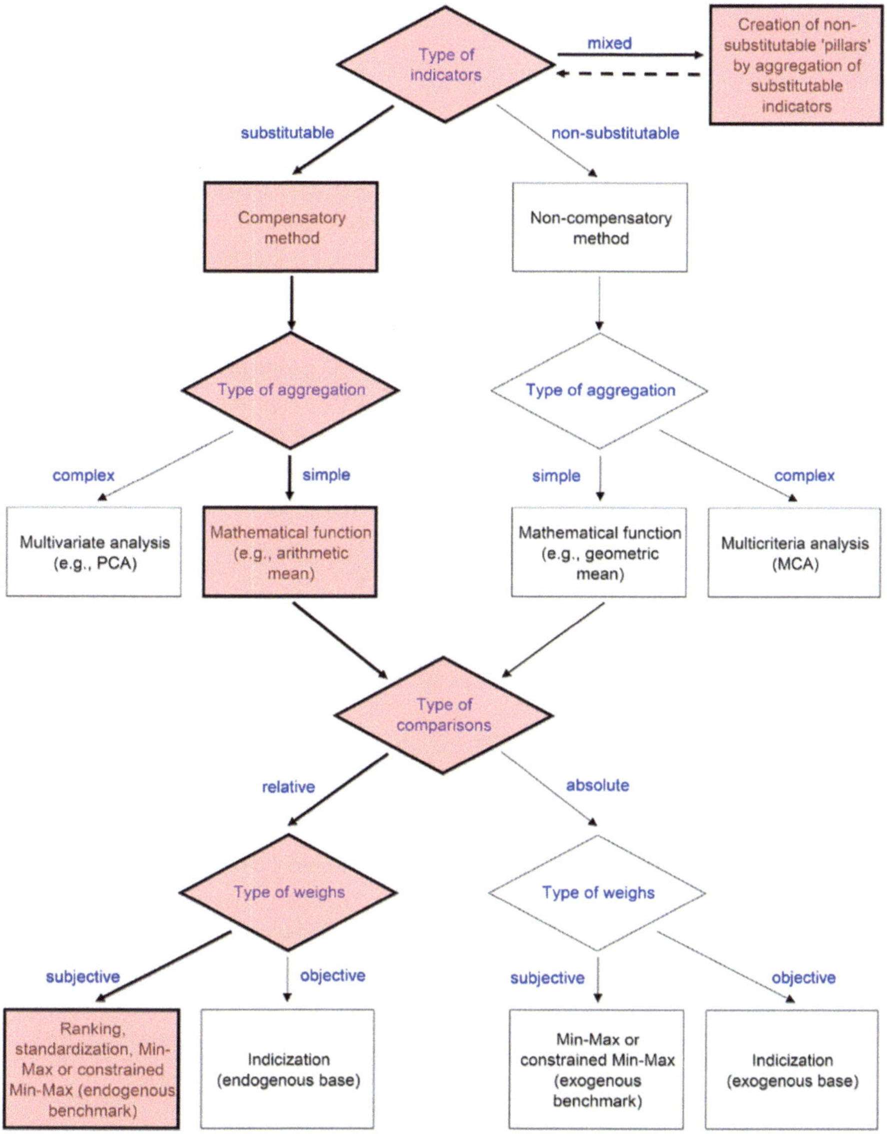

Fig. 5.3 Flowchart for constructing the RCI

5.6 Digital Economy and Society Index (DESI)

The *Digital Economy and Society Index* (DESI) summarizes indicators on Europe's digital performance and tracks the progress of EU countries.

Table 5.6 Individual indicators of the EPI (Year 2022)

Policy objective	Issue category	Indicator
Climate change	Climate change mitigation	Projected GHG emissions in 2050
		CO_2 growth rate
		CH_4 growth rate
		CO_2 from land cover
		GHG intensity trend
		F-gas growth rate
		Black carbon growth rate
		GHG emissions per capita
		N_2O growth rate
Environmental health	Air quality	PM2.5 exposure
		Household solid fuels
		Ozone exposure
		NOx exposure
		SO_2 exposure
		CO Exposure
		VOC exposure
	Sanitation and drinking water	Unsafe drinking water
		Unsafe sanitation
	Heavy metals	Lead exposure
	Waste management	Mismanaged solid waste
		Recycling rates
		Ocean plastic pollution
Ecosystem vitality	Biodiversity and habitat	Terrestrial biome protection (national)
		Terrestrial biome protection (global)
		Marine protected areas
		Protected Areas Rep. Index
		Species habitat index
		Species protection index
		Biodiversity habitat index
	Ecosystem services	Tree cover loss
		Grassland loss
		Wetland loss
	Fisheries	Fish stock status
		Marine trophic index
		Fish caught by trawling
	Acid rain	SO_2 growth rate

(continued)

Table 5.6 (continued)

Policy objective	Issue category	Indicator
		NOx growth rate
	Agriculture	Sustainable nitrogen Mgmt. Index
		Sustainable pesticide use
	Water resources	Wastewater treatment

Source Wolf et al. (2022)

Since 2014, the European Commission has been monitoring the digital progress of member states using this index, which identifies areas where priority action is needed. The reports, published annually, describe the profiles of each Member State and provide an analysis of the key areas of digital policy at European level. Over the years, the European Commission has modified the DESI to align it with the four points of the Commission's proposal "Path to the Digital Decade", which sets out the concrete digital targets that the Union as a whole should achieve by 2030 for a complete digital transformation in all sectors of the economy. In this context, eleven indicators of the DESI 2022 measure the Digital Decade targets.

The DESI covers four dimensions: human capital, connectivity, integration of digital technology and digital public services. The dimensions relate to the four most important policy areas of the "Digital Compass 2030". As these are interlinked areas, the development of the digital economy and society cannot be achieved through isolated improvements in specific sectors, but only through concerted progress in all sectors.

Indicators Selection
The individual indicators were selected on the basis of the following criteria:

- They must be collected regularly. In order to fulfil the monitoring function, the indicators used in the index must ideally be collected annually (or at least with a predefined periodicity).
- They must be relevant to a policy area of interest. All indicators in the index must be recognized as relevant metrics in the respective policy areas.
- They must not be redundant.

The indicators used in the DESI 2022 are listed in Table 5.7.

Normalization
The indicators are normalized using the Min–Max method to convert all values to [0, 1].

In order to enable comparisons of the index scores over time, the minima and maxima for the normalization of each indicator were defined and will also be used for normalization in the future versions of the DESI (European Commission 2022).

Table 5.7 Individual indicators of the DESI (Year 2022)

Dimension	Sub-dimension	Indicator
Human capital	Internet user skills	At least basic digital skills
		Above basic digital skills
		At least basic digital content creation skills
	Advanced skills and development	ICT specialists
		Female ICT specialists
		Enterprises providing ICT training
		ICT graduates
Connectivity	Fixed broadband take-up	Overall fixed broadband take-up
		At least 100 Mbps fixed broadband take-up
		At least 1 Gbps take-up
	Fixed broadband coverage	Fast broadband (NGA) coverage
		Fixed Very High Capacity Network (VHCN) coverage
	Mobile broadband	5G spectrum
		5G coverage
		Mobile broadband take-up
	Broadband prices	Broadband price index
Integration of digital technology	Digital intensity	SMEs with at least a basic level of digital intensity
	Digital technologies for businesses	Electronic information sharing
		Social media
		Big data
		Cloud
		AI
		ICT for environmental sustainability
		e-Invoices
	e-Commerce	SMEs selling online
		e-Commerce turnover
		Selling online cross-border
Digital public services	e-Government	e-Government users
		Pre-filled forms
		Digital public services for citizens

(continued)

Table 5.7 (continued)

Dimension	Sub-dimension	Indicator
		Digital public services for businesses
		Open data

Aggregation

The normalized indicators are aggregated into sub-dimensions, the sub-dimensions into dimensions and the dimensions into the overall index using weighted arithmetic means. The formula for calculating the DESI is as follows:

$$\begin{aligned} \mathrm{DESI} = {} & \mathrm{I}_{\mathrm{Human\_capital}} \cdot 0.25 + \mathrm{I}_{\mathrm{Connectivity}} \cdot 0.25 \\ & + \mathrm{I}_{\mathrm{Integration\_of\_digital\_technology}} \cdot 0.25 + \mathrm{I}_{\mathrm{Digital\_public\_services}} \cdot 0.25. \end{aligned}$$

where $\mathrm{I}_{\mathrm{Human\_capital}}$, $\mathrm{I}_{\mathrm{Connectivity}}$, $\mathrm{I}_{\mathrm{Integration\_of\_digital\_technology}}$ and $\mathrm{I}_{\mathrm{Digital\_public\_services}}$ are the partial composite indices of the four dimensions.

The scheme for constructing the DESI is shown in Fig. 5.4.

5.7 Other Indices 'Beyond GDP'

Personal Well-Being Index (PWI)

The *Personal Well-being Index* (PWI) was developed to measure subjective well-being (SWB), also known as subjective quality of life (SQOL). SWB provides a comprehensive, global and complete view of quality of life, usually measured through life satisfaction questions in surveys of individuals.

In 2002, the International Well-being Group was established (Cummins and Lau 2003) with the aim of developing a survey instrument for SWB. The scale for measuring PWI is a further development of the comprehensive quality of life scale (ComQol), which was originally developed by Cummins (1997). The ComQol included both an objective and a subjective measure of quality of life, whereas the PWI only included questions on satisfaction. In addition, the format of the response scale was changed, in particular the original 7-point Likert scale was replaced by an 11-point scale (0–10).

The PWI is currently used in regular surveys of the Australian population to measure satisfaction with the following areas: standard of living, personal health, achieving in life, personal relationships, personal safety, community-connectedness, and future security. These seven domains are theoretically linked as they stem from the first question about satisfaction with life as a whole (International Wellbeing Group 2013). The scores obtained in each domain are added together to obtain an average satisfaction score representing SWB (Lau et al. 2005).

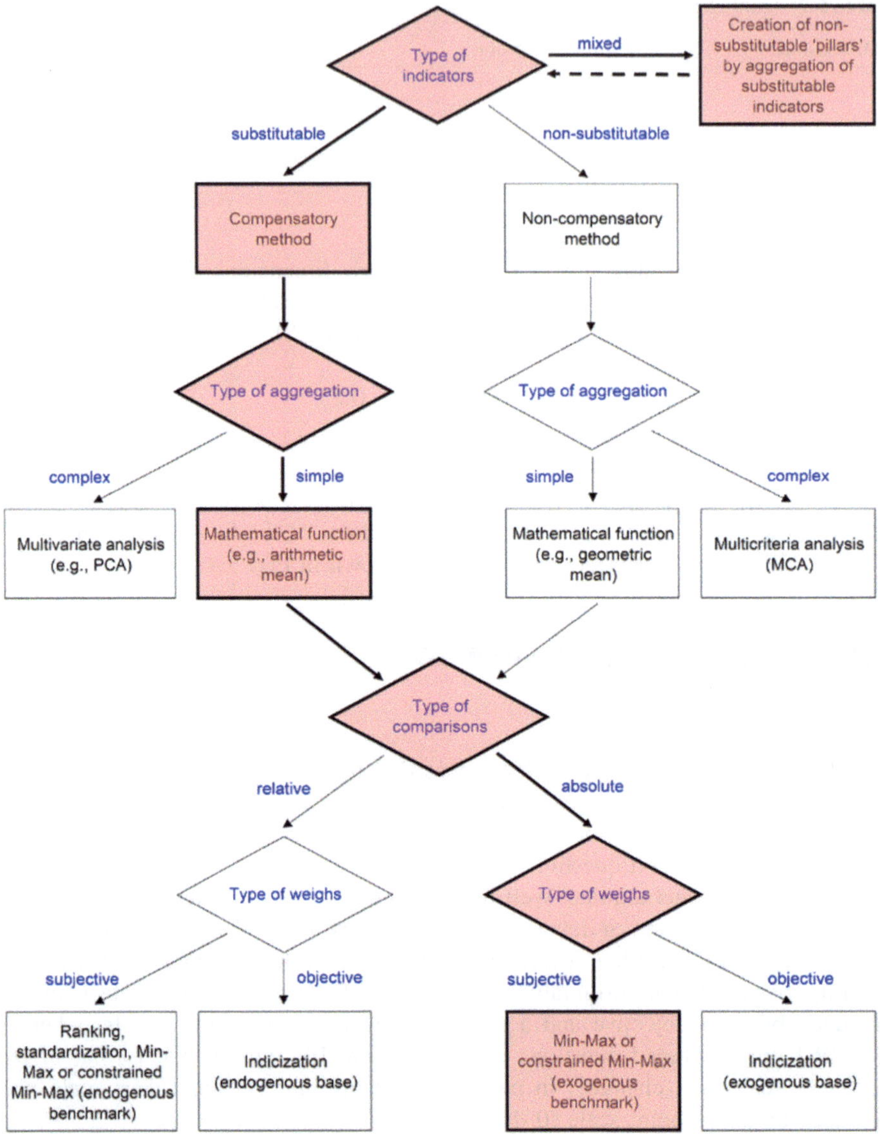

Fig. 5.4 Flowchart for constructing the DESI

Psychometric analyses conducted to test the validity, reliability and sensitivity of the PWI have yielded positive results in the Australian context (Cummins et al. 2003).

Happy Planet Index (HPI)
The *Happy Planet Index* (HPI) is a measure of sustainable well-being developed by the New Economics Foundation in 2006. It is the first index that combines environmental impact with well-being, rating countries on how efficiently they enable their people to live long, happy lives using our limited environmental resources. The HPI reflects the average years of 'happy life' that a given country produces per unit of planetary resources consumed. In other words, it represents the efficiency with which countries convert the Earth's limited resources into the well-being of their citizens.

The composite index is based on three indicators: "Life expectancy" ($I_{Life\_expectancy}$), "Self-reported well-being" ($I_{Self-reported\_well-being}$) and "Carbon footprint" ($I_{Carbon\_footprint}$).

Life expectancy is from the UN Population Division, self-reported well-being is measured using a question from Gallup World Poll on a scale of 0–10 and carbon footprint is an estimate of per capita greenhouse gas emissions associated with consumption and economic activity in a country (Abdallah et al. 2024).

The HPI is calculated using the following formula[3]:

$$\mathrm{HPI} = \left(I_{Life\_expectancy} \cdot I_{Self-reported\_well-being}\right)/I_{Carbon\_footprint}$$

and can be seen as a measure of 'well-being per tonne of CO_2 emissions'.

The countries that perform best have a relatively high life expectancy and a high self-reported well-being with a relatively small carbon footprint. Therefore, the HPI not only measures well-being, but also sustainable well-being.

References

Abdallah S, Hoffman A, Akenji L (2024) The 2024 happy planet index. Hot or Cool Institute, Berlin

Annoni P, Kozovska K (2010) EU Regional Competitiveness Index (RCI) 2010. Publications Office of the EU, Luxembourg

Canadian Index of Wellbeing (2012) How are Canadians really doing? The 2012 CIW report. Waterloo

Cummins RA (1997) Comprehensive quality of life scale—adult, manual, 5th edn. School of Psychology. Deakin University, Melbourne

Cummins RA, Lau ALD (2003) An introduction to the international wellbeing group and the international wellbeing index. In: Fifth conference of the international society for quality-of-life studies, Frankfurt, Germany

Cummins RA, Eckersley R, Pallant J, Van Vugt J, Misajon R (2003) Developing a national index of subjective wellbeing: the Australian unity wellbeing index. Soc Indic Res 64:159–190

Dijkstra L, Papadimitriou E, Cabeza Martinez B, De Dominicis L, Kovacic M (2023) EU regional competitiveness index 2.0. Working Paper, No 1, DG Regional and Urban Policy

[3] The equation includes some statistical adjustments, which are described in detail in the methodological note (Hot or Cool Institute, 2024).

Dijkstra L, Annoni P, Kozovska K (2011) A new regional competitiveness index: theory, methods and findings. Working Paper, No 2, DG Regional and Urban Policy
European Commission (2022) Digital economy and society index (DESI) 2022 methodological note. https://ec.europa.eu/newsroom/dae/redirection/document/88557
Hall J, Giovannini E, Morrone A, Ranuzzi G (2010) A Framework to measure the progress of societies. OECD Statistics Working Papers 5, OECD Publishing
Hot or Cool Institute (2024) The 2024 happy planet index. Methodology paper. https://happyplanetindex.org/HPI_2024_methodology.pdf
International Wellbeing Group (2013) Personal wellbeing index, 5th ed. Australian Centre on Quality of Life, Deakin University, Melbourne
Istat (2015) BES 2015. Equitable and sustainable well-being in Italy. Istituto nazionale di statistica, Rome
Istat (2023) BES 2022. Equitable and sustainable well-being in Italy. Istituto nazionale di statistica, Rome
Kahneman D, Deaton A (2010) High income improves evaluation of life but not emotional well-being. Proc Natl Acad Sci 107(38):16489–16493
Lau ALD, Cummins RA, McPherson W (2005) An investigation into the cross-cultural equivalence of the personal wellbeing index. Soc Indic Res 77:403–430
Michalos AC, Smale B, Labonté R, Muharjarine N, Scott K et al (2011) The Canadian index of wellbeing, technical report 1.0. Canadian Index of Wellbeing and University of Waterloo, Waterloo, ON
Schwab K (ed) (2012) The global competitiveness report 2012–2013. World Economic Forum, Geneva
Schwab K, Porter ME (2007) The global competitiveness report 2007–2008. World Economic Forum, Geneva
Wendling ZA, Emerson JW, de Sherbinin A, Esty DC et al (2020) 2020 Environmental performance index. Yale Center for Environmental Law & Policy, New Haven, CT
Wolf MJ, Emerson JW, Esty DC, de Sherbinin A, Wendling ZA et al (2022) 2022 Environmental performance index. Yale Center for Environmental Law & Policy, New Haven, CT

MIX
Papier aus verantwortungsvollen Quellen
Paper from responsible sources
FSC® C105338

If you have any concerns about our products,
you can contact us on
ProductSafety@springernature.com

In case Publisher is established outside the EU,
the EU authorized representative is:
Springer Nature Customer Service Center GmbH
Europaplatz 3, 69115 Heidelberg, Germany

Printed by Libri Plureos GmbH
in Hamburg, Germany